Water Activity
and Food

FOOD SCIENCE AND TECHNOLOGY

A SERIES OF MONOGRAPHS

Maynard A. Amerine, Rose Marie Pangborn, and Edward B. Roessler, PRINCIPLES OF SENSORY EVALUATION OF FOOD. 1965.

C. R. Stumbo, THERMOBACTERIOLOGY IN FOOD PROCESSING, second edition. 1973.

Gerald Reed (ed.), ENZYMES IN FOOD PROCESSING, second edition. 1975.

S. M. Herschdoerfer, QUALITY CONTROL IN THE FOOD INDUSTRY. Volume I – 1967. Volume II – 1968. Volume III – 1972.

Hans Riemann, FOOD-BORNE INFECTIONS AND INTOXICATIONS. 1969.

Irvin E. Liener, TOXIC CONSTITUENTS OF PLANT FOODSTUFFS. 1969.

Martin Glicksman, GUM TECHNOLOGY IN THE FOOD INDUSTRY. 1970.

L. A. Goldblatt, AFLATOXIN. 1970.

Maynard A. Joslyn, METHODS IN FOOD ANALYSIS, second edition. 1970.

A. C. Hulme (ed.), THE BIOCHEMISTRY OF FRUITS AND THEIR PRODUCTS. Volume 1 – 1970. Volume 2 – 1971.

G. Ohloff and A. F. Thomas, GUSTATION AND OLFACTION. 1971.

George F. Stewart and Maynard A. Amerine, INTRODUCTION TO FOOD SCIENCE AND TECHNOLOGY. 1973.

Irvin E. Liener (ed.), TOXIC CONSTITUENTS OF ANIMAL FOODSTUFFS. 1974.

Aaron M. Altschul (ed.), NEW PROTEIN FOODS: Volume 1, TECHNOLOGY, PART A – 1974. Volume 2, TECHNOLOGY, PART B – 1976.

S. A. Goldblith, L. Rey, and W. W. Rothmayr, FREEZE DRYING AND ADVANCED FOOD TECHNOLOGY. 1975.

R. B. Duckworth (ed.), WATER RELATIONS OF FOOD. 1975.

A. G. Ward and A. Courts (eds.), THE SCIENCE AND TECHNOLOGY OF GELATIN. 1976.

John A. Troller and J. H. B. Christian, WATER ACTIVITY AND FOOD. 1978.

In preparation

D. R. Osborne and P. Voogt, THE ANALYSIS OF NUTRIENTS IN FOODS.

A. E. Bender, FOOD PROCESSING AND NUTRITION.

Water Activity and Food

JOHN A. TROLLER

Winton Hill Technical Center
The Procter & Gamble Company
Cincinnati, Ohio

J. H. B. CHRISTIAN

Commonwealth Scientific and Industrial
Research Organization
Division of Food Research
North Ryde, New South Wales, Australia

ACADEMIC PRESS
New York San Francisco London 1978
A Subsidiary of Harcourt Brace Jovanovich, Publishers

ACADEMIC PRESS, INC.
111 Fifth Avenue, New York, New York 10003

United Kingdom Edition published by
ACADEMIC PRESS, INC. (LONDON) LTD.
24/28 Oval Road, London NW1 7DX

Library of Congress Cataloging in Publication Data

Troller, John A
Water activity and food.

(Food science and technology)
Includes bibliographies and index.
1. Food--Water activity. 2. Food--Microbiology.
3. Food industry and trade. I. Christian, J. H. B., joint author. II. Title.
TX553.W3T76 664'.02 77-11226
ISBN 0-12-700650-8

PRINTED IN THE UNITED STATES OF AMERICA

To W. C. FRAZIER
and W. J. SCOTT

Pioneer researchers in food microbiology and microbial water relations.

Contents

4 LIPID OXIDATION, CHANGES IN TEXTURE, COLOR, AND NUTRITIONAL QUALITY

5 MICROBIAL GROWTH

6 FOOD PRESERVATION AND SPOILAGE

7 MICROBIAL SURVIVAL

8 FOOD-BORNE PATHOGENS

Foreword

It is not yet 200 years since Cavendish (1781) discovered that water was a compound produced when hydrogen was burned in oxygen. It is also less than 100 years since van't Hoff (1887) announced the relationship between osmotic pressure and the molecular weight of solutes. In the same year (1887–1888) Raoult showed that the molecular weight of solutes could also be determined by measuring the vapor pressure of solutions. Against these landmarks in the progress of science, we can note that it is only during the last 20 years that the use of water activity to describe the water status of the environment has gained increased acceptance by microbiologists and, more recently, by food scientists and technologists. And now we have a book, "Water Activity and Food," by two authors who are both very well qualified to write on water activity as a factor in the water relations of microorganisms, and in food processing and storage.

After an introductory chapter on water in foods and in solutions, and a second chapter on methods, there follow two chapters dealing with the water relations of enzyme activity, lipid oxidation, nonenzymatic browning, and several other food-related factors. The water relations of microbial growth, the effects of water on microbial survival, the spoilage and preservation of foods at various levels of water activity, and the water relations of food-borne pathogens are next discussed in some detail. The final three chapters deal with the importance of water activity in nonmicrobiological aspects of food processing and storage. Throughout the book, the authors examine the literature from an "a_w point of view," and give numerous examples of the value of water activity as a basis for predicting the reactions of microorganisms or the stability of food components. They also report on some examples where water activity has been a somewhat inadequate predictor of events, and on a number of interesting interactions with other environmental parameters.

No matter how useful the concept of water activity may be in certain applications in food science and technology, it should not be expected to explain everything. Water activity is, after all, simply a number which provides information about the vapor pressure of water in a system. While it describes the exact percentage of water molecules in a system which has the characteristics of pure water, it tells us nothing of the nature of the forces binding the remainder. It is, in fact, nothing more than a statement of the average properties of all the water molecules in the system. A proper understanding of the role of water in complex systems, such as living cells and foods, must depend on much more detailed information about all the properties of the water molecule and its constituent ions, and of their interactions with the great variety of molecules and ions in the environment. The task is undoubtedly formidable but the steadily increasing array of physical methods for the study of such interactions will enable some progress to be made.

Addressed as it is to food scientists and technologists, this book should help to bring a more enlightened approach to studies of foods and their associated microorganisms, in relation to the water status of the environment. It should assist food technologists interested in developing new products and processes to make the best use of present knowledge about the significance of water in foods. It should also stimulate other investigators to measure and control water activity in experiments in which the role of water is being studied. In this way, we shall be able to fill in some of the obvious gaps in our knowledge and will soon discover whether a_w can be replaced by something more meaningful or whether, like pH, it will be something to measure and consider for perhaps another 100 years or so.

W. J. SCOTT

Preface

Throughout the past twenty-five years, the importance of water activity in foods has received emphasis in the work of food microbiologists. Over a similar period, chemists have likewise paid attention to its influence in enzymatic and non-enzymatic reactions in foods. Physicists, too, have long considered the effects of relative humidity, water activity, and other manifestations of water vapor pressure, particularly as these relate to matters of heat and mass transfer in foods. Indeed, in all of these disciplines, there exist specialized reviews of water activity and its significance in foods.

However, it is the authors' impression that only relatively recently—since the discussions on intermediate moisture foods—has the food technologist become fully aware of the importance of water activity in both food preservation and processing. Our intention, therefore, is to give the food technologist an overview of the influence of water activity on the systems with which he is concerned. Accordingly, we have omitted both detailed considerations of the theory of water structure, and in-depth discussions of the physiological basis of microbial water relations. It is our hope that this book, while being of assistance to food technologists, may permit those in specialist disciplines to appreciate the significance of water activity in areas of food science outside their own fields.

The book is organized to proceed from the particular to the general, that is, from physical and chemical aspects to microbiology, food processing and storage, and sanitation. At the risk of some overlap, we have included a section on pathogenic microorganisms separate from that dealing with general microbiological considerations. It is felt that many readers will appreciate this special treatment.

In attempting a broad coverage of this topic, we have relied heavily on the advice of colleagues. Scientists of both the CSIRO Division of Food Research and The Procter & Gamble Company Winton Hill Technical Center have been most helpful with comment and criticism. We also are especially indebted to A. C. Baird-Parker, T. P. Labuza, W. J. Scott, and J. G. Voss, all of whom contributed invaluable suggestions and advice. However, the responsibility for the text rests solely with the authors who conceived and wrote it. Finally, we acknowledge, with thanks, our wives, Betty and Jacky, who, although a hemisphere apart, shared many of the concerns and frustrations attendant to the genesis of this volume.

JOHN A. TROLLER
J. H. B. CHRISTIAN

Water Activity and Food

1

Water Activity—Basic Concepts

WATER IN FOODS

All foods contain water, and it is a common observation that the foods most likely to show rapid deterioration due to biological and chemical changes are usually those of high water content. The application of this principle precedes recorded history, but at some stage, our early ancestors found that spoilage could be delayed or prevented by using the available means to dry perishable foods, particularly the flesh of beast, fish, and fowl. The heat of fires, the sun's radiation, and even the sublimation of ice at high altitudes or high latitudes produced items of diet that could be kept for later consumption without many of the adverse changes apparent when the untreated material was so stored. Smoking, presumably an accidental adjunct to the drying of flesh over fires, added antimicrobial substances that augmented the stabilizing effect of drying.

The water in foods serves as solvent for many constituents, and any drying process will concentrate these solutions. If it is increasing concentration, rather than decreasing water content per se, that preserves food, other methods of increasing the concentration of a food's aqueous phase should also enhance its stability. Such alternatives as salting and syruping do preserve foods by increasing solute concentrations. The salting of pork for shipboard use in the days of sail is a classic example. Historically, the combination of salting and drying was used commonly for foods of animal origin, and, later, syruping and drying were utilized for the preservation of fruits.

Thus, the study of water in foods is in large part a study of aqueous solutions in which the solutes, by dint of both their nature and concentration, alter the physical properties of the solvent. Removing water from foods and adding solutes to foods have demonstrably similar, but rarely identical effects upon, for example, their resistance to microbial attack. When the food solution has been concentrated to the point at which microbial growth is controlled, susceptibility to certain undesirable chemical and physical changes becomes apparent. Greater increases in concentration may be required to control these enzymatic and nonenzymatic changes which influence, in particular, the taste and appearance of food.

The water-related criteria that have been used in studies of the stability of foods include water content, solute concentration, osmotic pressure, equilibrium relative humidity (E.R.H.), and water activity (a_w). Water content and solute concentration tell little about the properties of the water present in the food. However, the other indices are measurements of related colligative or osmotic properties and appear to be good indicators of the availability of the water to participate in reactions. All are not equally appropriate to food—the use of osmotic pressure presupposes the presence of a membrane of suitable permeability, and equilibrium relative humidity refers strictly to the atmosphere in equilibrium with a food, not to the food itself.

This book is concerned with the remaining parameter, a_w. There is now wide agreement that a_w is the most useful expression of the water requirements for, or water relations of, microbial growth (Scott, 1957) and enzyme activity (Acker, 1962). The alternatives of solute concentration and water content have been shown very clearly by Scott (1962) to be inadequate for describing the availability of water for the multiplication of certain bacteria (see Chapter 5, Table 5.2).

The expressions referred to above all relate to equilibrium conditions. However, in many situations of concern in food technology, the water in a food is not in equilibrium with the water vapor in its surroundings. In such cases, the water content of the food will change with time, and this may have profound effects on the stability of the food. These considerations are particularly relevant to the packaging and storage of foods and will be discussed in later chapters.

PROPERTIES OF SOLUTIONS

When solutes are dissolved in water, entropy is decreased as water molecules become oriented with respect to solute molecules. Water molecules are less free to escape from the liquid into the vapor phase, and the vapor pressure is lowered. Consequent upon this are the depression of freezing point and the elevation of boiling point. The relationship between concentration and vapor pressure for ideal solutions is given by Raoult's law. This law states that for ideal solutes, the relative lowering of the vapor pressure of the solvent is equal to the mole fraction of the solute.

If p and p_0 are the vapor pressures of solution and solvent, respectively, and if

n_1 and n_2 are the number of moles of solute and solvent, respectively, Raoult's law may be expressed

$$\frac{p_o - p}{p_o} = \frac{n_1}{n_1 + n_2} \tag{1}$$

As 1 kg of water contains 55.51 moles, 1 mole of an ideal solute dissolved in 1 kg of water will lower the vapor pressure by 1/(1 + 55.51) or 0.0177, i.e., by 1.77%. More conveniently, Raoult's law may be written

$$\frac{p}{p_o} = \frac{n_2}{n_1 + n_2} \tag{2}$$

indicating that, for the 1 molal solution, the vapor pressure is 55.51/(1 + 55.51) or 0.9823 or 98.23% of that of pure water. The ratio of the vapor pressures of solution and solvent is described by the term water activity (a_w),

$$a_w = \frac{p}{p_o} \tag{3}$$

so that the 1 molal solution referred to above has an a_w of 0.9823. An atmosphere in water vapor equilibrium with this solution will have a relative humidity of 98.23%. Thus, under equilibrium conditions, E.R.H. is equal to $a_w \times 100$. Provided that its vapor pressure is not reduced by interaction with insoluble materials, pure water has an a_w of 1.00, which is equivalent to an E.R.H. of 100%. The influence of water binding on a_w is discussed below.

The solutes of concern in foods are not ideal, and many will cause much greater depressions of vapor pressure and, hence, a_w than are predicted by Raoult's law. For nonelectrolytes, the difference may be small, at least up to 1 molal concentrations, but for electrolytes the effect is always great, increasing

TABLE 1.1.

Values of a_w for Aqueous Solutions of 1 Molal Concentration

Solute	a_w
Ideal solute	0.9823
Glycerol	0.9816
Sucrose	0.9806
Sodium chloride	0.967
Calcium chloride	0.945

with the increase in the number of ions generated per molecule. Compare the a_w values for 1 molal solutions listed in Table 1.1. Thus, to calculate a_w values for nonideal solutes, a molal osmotic coefficient (ϕ) is used in the formula

$$\log_e a_w = \frac{-vm\ \phi}{55.51} \tag{4}$$

v is the number of ions generated by each molecule of solute and is equal to 1 for nonelectrolytes; m is the molal concentration. Values for ϕ are given by Robinson and Stokes (1955) for electrolytes, by Scatchard *et al.* (1938) for nonelectrolytes, and by Wolf (1966). Scott (1957) discusses the use of this formula. Tables of a_w and concentration for a range of solutes are given in Chapter 2.

Relative humidity has been widely used in mycology in the past to define the water relations of fungi, while physiologists have generally made reference to osmotic pressures in relevant studies of both plant and animal cells. As noted previously, these are less appropriate than a_w for the purpose. The relationship of a_w to E.R.H. has been mentioned; osmotic pressure may be converted to a_w by the expression

$$\text{Osmotic pressure} = \frac{-RT}{\bar{V}} \log_e a_w \tag{5}$$

where $\bar{V}$ is the partial molar volume of water.

In such pure solutions as have been discussed above, the a_w differs from that predicted by Raoult's law primarily as a consequence of dissociation, where this occurs, and of interactions between solute molecules. In foods, as Karel (1973) points out, the situation is more complicated. Not all of the water is free to act as solvent, some being bound to specific groups of insoluble components. Additionally, some of the solutes also are bound to insoluble components. Thus, a knowledge of the water content of a food and the nature and amount of solutes present does not provide a basis for an accurate calculation of the food's a_w level.

WATER BINDING

The depression of water vapor pressure and, hence, a_w by the binding of water molecules in foods has received much attention. The most popular approach has been the Brunauer–Emmett–Teller (BET) isotherm of Brunauer *et al.* (1938), which, although derived from observations on nonpolar gases, gives very useful estimates of the size of the monolayer of adsorbed water molecules. Another approach utilizes measurements of the total heat of absorption of water as a function of a_w (Karel, 1973). Both techniques give indications of the level of hydration at which the mobility of small molecules becomes apparent and of the total number of polar groups binding water. However, the greatest value in knowing the a_w at which the monolayer exists is that it appears to be the most stable water content for most foods. Lipid oxidation rates increase at water contents below the monolayer, while rates of nonenzymatic browning increase above it (Labuza, 1968).

The a_w may also be depressed by the capillary forces generated by water held within capillaries in foods. While knowledge of the dimensions of capillaries actually present in foods is meager, Karel (1973) suggests that capillaries with a radius of 10^{-6} cm are probably common, which would lead to definite capillary effects above an a_w level of 0.90. The capillarity of foods appears to play an important role also in the phenomenon of hysteresis, as will be discussed later.

WATER SORPTION ISOTHERMS

It will now be clear that the a_w of a food containing substantial amounts of insoluble constituents cannot be calculated accurately from a knowledge of the concentrations of the component solutes. It is necessary, therefore, to measure the a_w directly by methods outlined in Chapter 2. To obtain comprehensive data on the water relations of a food, the a_w levels corresponding to a range of water contents must be determined. These are plotted to provide a water sorption isotherm. This isotherm is useful, not only in showing at what water contents certain desirable or undesirable levels of a_w are achieved, but also in indicating what significance particular changes in water content will have in terms of a_w.

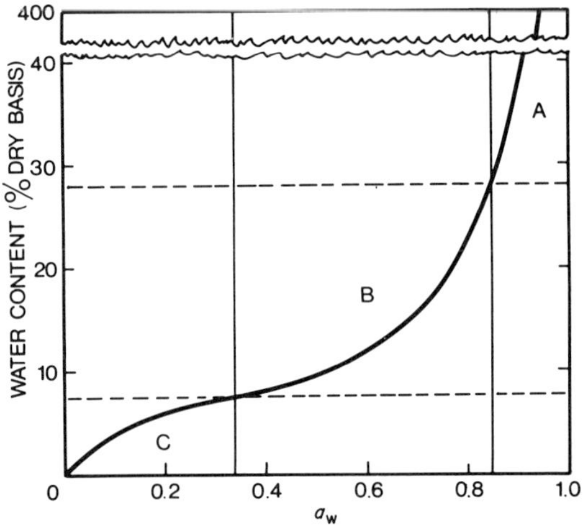

Fig. 1.1. A generalized water sorption isotherm of a typical food material. Regions A, B, and C are discussed in the text (Duckworth, 1974).

The schematic isotherm for a typical food material shown in Fig. 1.1 is taken from Duckworth (1974). The forces responsible for reducing the water vapor pressure are not the same throughout the range of a_w values, and A, B, and C on this isotherm represent regions in which different types of water binding may predominate.

As water is added to a dry food material, molecules are adsorbed onto appropriate sites until, statistically at least, all are occupied. This constitutes the water monolayer, formed in region C of the isotherm (Fig. 1.1). It is near the point of monolayer completion that given changes in water content have the most marked influence upon a_w. Conversely, relatively large changes in the a_w of the food are necessary to cause appreciable evaporation or condensation of water. The water of the monolayer (region C, Fig. 1.1) is thus very stable, behaving in many ways as part of the solid, and believed to be nonfreezable at any temperature (Duckworth, 1974). Foods that consist largely of soluble, low molecular weight components may have such a small monolayer that the relationship shown in region C of Fig. 1.1 does not occur.

The water in region B of the isotherm (Fig. 1.1) is less firmly bound than in the monolayer. Multilayer adsorption occurs, and the solution of soluble components becomes important, modified by the nature of insoluble solids present. The water in regions B and C differs markedly from "free" water, such as exists in region A. This latter, although mechanically trapped in the system, is subjected to only weak restrictive forces, as indicated by the steepness of the isotherm.

TEMPERATURE EFFECTS

Note that the graphical relationships between water content and a_w are termed "isotherms." Changes in temperature affect the relation markedly, as is shown by the isotherms of raw chicken in Fig. 1.2 (Wolf *et al.*, 1973). Moreover, temperature gradients in a food material lead to water vapor pressure gradients, with resulting transfer of moisture and changes in a_w levels. The importance of such changes to the storage of foodstuffs is discussed in Chapter 10.

Many attempts have been made to formulate equations that will describe the temperature dependency of the isotherm. Most are very complex, involving the use of four constants. There is a simpler equation based on only two parameters (Iglesias and Chirife, 1976), which may be of more practical value, although it gives no indication of the physicochemical mechanism of water sorption.

HYSTERESIS

Temperature is not the only variable that can influence the amount of sorbed water in a food sample at a given a_w level. The water content will, for many

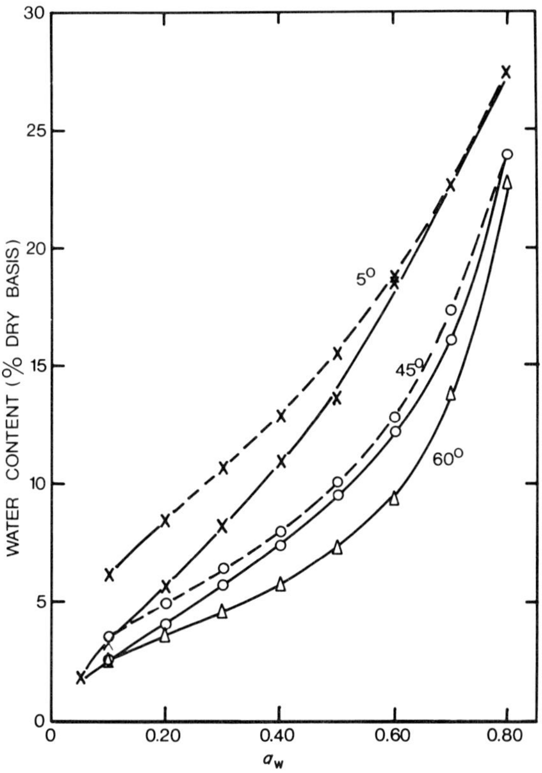

Fig. 1.2. Adsorption (——) and desorption (---) isotherms for raw chicken at three temperatures. Adsorption and desorption data were identical at 60°C (Wolf *et al.*, 1973).

foods, be higher when the a_w is achieved by desorption of water from a moist material than when the route is by adsorption to a dry food. This difference is termed hysteresis and is illustrated in Fig. 1.2 for raw chicken, in which it is greatest at 5°C and not detectable at 60°C.

A general sorption isotherm with regions as discussed by Labuza (1968) is shown in Fig. 1.3. This exaggerated hysteresis loop terminates near the a_w of monolayer formation, but, as mentioned earlier, the monolayer region A is not evident in all foods. Here region B is much narrower than region B of Fig. 1.1, and Labuza (1968) ascribes it to the adsorption of further layers of water on the monolayer. Region C of Fig. 1.3 is ascribed to condensation in pores (capillary effects), followed by dissolution of soluble components. This suggests that in this example hysteresis is a consequence predominantly of capillary condensation.

Thus, in theory, the course of water sorption by a dry material is first by the formation of a monolayer, followed by multilayer adsorption, the uptake into

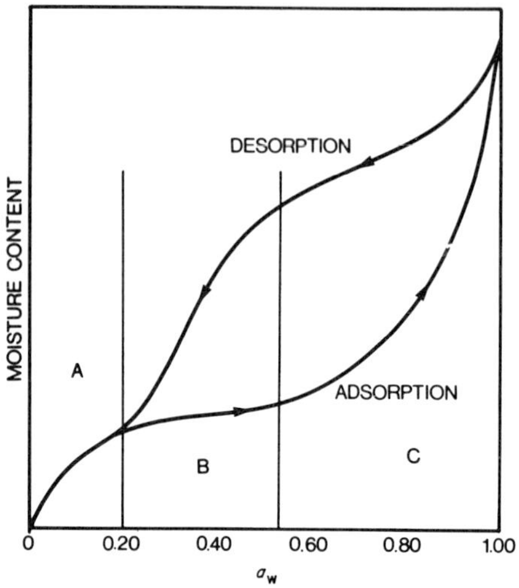

Fig. 1.3. Generalized water sorption isotherm showing a hysteresis loop (Labuza, 1968).

pores and capillary spaces, dissolution of solutes, and finally, mechanical entrapment of water at the higher levels of a_w. These phases may overlap extensively and will differ in magnitude among foods depending upon chemical composition and structure.

The extent and shape of the hysteresis loop also vary greatly among foods. Wolf *et al.* (1972) describe sorption curves and hysteresis loops for several types of foods. In the apple, a high pectin- and sugar-containing food, hysteresis occurs predominantly in the monolayer region and not at all above 0.65 a_w. For pork, a high protein food, hysteresis begins at about 0.85 a_w, suggesting capillary condensation as a major contributor, but with a maximum at 0.15 a_w in the monolayer region. A starchy food, rice, shows a relatively large hysteresis effect at all levels of a_w below about 0.85, but with a maximum at 0.65. These examples serve to illustrate the complexity of water sorption in foods and the difficulties in ascribing hysteresis to any particular food–water interaction.

In practice, a food being processed will usually arrive at a particular level of a_w by a standard procedure that may always be desorptive or may always be adsorptive. Thus, any hysteresis effect is not as noticeable as it appears to be from the type of desorption–adsorption data discussed above. Indeed, data for microbial growth on foodstuffs, to be discussed later, do not show many of the inconsistencies that would result if hysteresis were an important factor. In fact, microbial behavior in most foods is very close to that predicted from data obtained in microbiological media in which hysteresis is insignificant (Christian and Scott, 1953; Scott, 1953). It is in the formulation of intermediate moisture foods (see Chapter 9) that hysteresis is most likely to complicate the use of a_w data as an index of stability to microbial attack.

FROZEN FOODS

The ice that separates when a food or any aqueous solution is cooled to its freezing point and below exerts a vapor pressure which depends only on the temperature of the ice. At equilibrium, the vapor pressure of this ice equals that of the water in the unfrozen solution within the food. It follows that the vapor pressure of the frozen food, and, hence, its a_w value, is defined by the temperature. Conversely, the concentration of solutes, and thus the a_w of a solution or food, determine the temperature at which it freezes, and useful estimates of a_w are obtained from measurements of freezing points.

The relationship among temperature, vapor pressure, and a_w in ice–water systems is shown in Table 1.2. It will be obvious that any reactions occurring in most commercially stored frozen foods take place not only at low temperatures, but also at levels of a_w substantially below those existing in the same foods when unfrozen.

In a related manner, unprotected frozen foods, when stored at freezing temperatures, suffer an opaque surface discoloration termed freezer burn. This is simply a sublimation of surface moisture. The rate of sublimation at constant air velocity is a function of the difference between the equilibrium vapor pressure of the food and the partial pressure of the water vapor in the air. In addition to color changes, sublimation may also cause weight losses that can be economically significant. These losses can be prevented by assuring that the vapor pressure difference does not exist or is, at least, minimal. An increase in the relative humidity of the atmosphere surrounding a stored frozen food will achieve this end. This may be obtained either by injection of water into the atmosphere of the storage room or by enclosing the food in a water-impermeable membrane.

A detailed discussion of sorption isotherms at temperatures below the freezing point is found in MacKenzie (1975).

TABLE 1.2.

Vapor Pressures of Water and Ice at Various Temperatures[a]

Temperature (°C)	Liquid water (mm Hg)	Ice (mm Hg)	$a_w = \frac{p_{ice}}{p_{water}}$
0	4.579	4.579	1.00
−5	3.163	3.013	0.953
−10	2.149	1.950	0.907
−15	1.436	1.241	0.864
−20	0.943	0.776	0.823
−25	0.607	0.476	0.784
−30	0.383	0.286	0.75
−40	0.142	0.097	0.68
−50	0.048	0.030	0.62

[a] Wolf *et al.*, 1973.

NONEQUILIBRIUM CONDITIONS

Water activities of foods must be determined under equilibrium conditions, i.e., when the food and the atmosphere in contact with it are in vapor pressure equilibrium. (Techniques for making such measurements are discussed in Chapter 2.) However, such conditions normally exist only in closed systems. A change in the relative humidity of the surrounding atmosphere leads to a change in the a_w of the food, and the type of packaging is important in determining the rate of such changes (see Chapter 10). Gradients in a_w are also common, with surface layers of foods hydrating or dehydrating rapidly in response to changes in the ambient atmosphere, but with no immediate changes in the depths of the food. These gradients assume great importance in dehydration in such methods as freeze-drying, on the one hand, and in salting (curing) and syruping of foods, on the other.

The mixing of food components of differing a_w levels can cause problems in stability. Salwin and Slawson (1959) demonstrated the paradox that, when packaged together, a drier food of low water content may transfer moisture to a wetter food of higher water content. This will occur when the latter food has the lower a_w. The extent of moisture transfer and, hence, the final equilibrium a_w value can be altered by varying the relative amounts of the two food components.

APPLICATIONS

Throughout this volume, many elements of a_w theory and related factors are discussed. As noted earlier, the intent is to provide basic knowledge from which at least some food-related phenomena can be explained and anticipated. For example, the extent to which a_w might be expected to influence the oxidative stability of a fat-containing system is discussed in Chapter 4, as well as the influence of a_w on antioxidant systems that might be employed to control this oxidation. With this knowledge, more effective processes or products may be designed.

Water activity-related concepts can also be applied in many other ways, such as the establishment of appropriate a_w limits and standards to achieve specific objectives. A manufacturer who understands that a nonrefrigerated cake icing can be stabilized with respect to microbial growth at an a_w level of 0.65 may specify that this a_w is not to be exceeded. Samples of the finished product can be withdrawn from the production line and tested by the most suitable means (see Chapter 2). Sampling plans, statistically derived and analyzed, will be useful in establishing sample numbers per production lot, the sample size, and the a_w limits required to produce a necessary confidence limit or interval. Alternatively, continuous in-line monitoring of a_w in the process stream may be feasible and may anticipate product problems arising from incorrect a_w levels. Often, an unsatisfactory a_w can be corrected without costly interruptions of production.

Public regulatory agencies also have shown interest in establishing a_w limits or standards for certain foods. The U.S. Food and Drug Administration has proposals for setting maximal levels of 0.85 a_w for some low-acid canned foods, and 0.70 for tree nuts and peanuts. The a_w is also the basis of guidelines for pickled, fermented, and acidified foods. International agencies, such as the FAO/WHO Codex Alimentarius Committee, are also considering a_w-related standards. Implicit in the use of such standards is the need for reliable, accurate, and appropriate methods of a_w measurement.

The concept of food standards based on a_w is fundamentally sound wherever a clear need for such regulations can be demonstrated. The a_w limits presently being considered are related to the minimal levels permitting microbial growth, toxin formation, or spore germination because of the relevance of these factors to health or product shelf life. As knowledge increases about the effects of a_w on, for example, the nutritional composition of foods, a_w standards related to this and other factors may follow.

WATER ACTIVITY VALUES FOR FOODS

The ensuing chapters discuss the influence of the a_w level of foods on the important chemical and biological changes that can take place. Clearly, not all food of a particular type will be identical in composition and, thus, of identical a_w. However, as a general guide, representative values of a_w for a selection of foods and ingredients are given in Appendix A.

REFERENCES

Acker, L. (1962). Enzymic reactions in foods of low moisture content. *Adv. Food Res.* **11,** 263–330.

Brunauer, S., Emmett, P. H., and Teller, E. (1938). Adsorption of gases in multimolecular layers. *J. Am. Chem. Soc.* **60,** 309–319.

Christian, J. H. B., and Scott, W. J. (1953). Water relations of salmonellae at 30°C. *Aust. J. Biol. Sci.* **8,** 75–82.

Duckworth, R. B. (1974). Water relationships of foods. *IFST (UK) Mini Symp. Dehydration, 1974,* pp. 6–9.

Iglesias, H. A., and Chirife, U. (1976). Prediction of the effect of temperature on water sorption isotherms of food material. *J. Food Technol.* **11,** 109–116.

Karel, M. (1973). Recent research and development in the field of low-moisture and intermediate moisture foods. *Crit. Rev. Food Technol.* **3,** 329–373.

Labuza, T. P. (1968). Sorption phenomena in foods. *Food Technol. (Chicago)* **22,** 263–272.

MacKenzie, A. P. (1975). The physico-chemical environment during the freezing and thawing of biological materials. *In* "Water Relations of Foods" (R. B. Duckworth, ed.), pp. 477–503. Academic Press, New York.

Robinson, R. A., and Stokes, R. H. (1955). "Electrolyte Solutions." Butterworth, London.

Salwin, H., and Slawson, V. (1959). Moisture transfer in combinations of dehydrated foods. *Food Technol. (Chicago)* **13,** 715–718.

Scatchard, G., Harman, W. J., and Wood, S. E. (1938). Isotonic solutions. I. The chemical potential of water in aqueous solutions of sodium chloride, potassium chloride, sulfuric acid, sucrose, urea and glycerol at 25°. *J. Am. Chem. Soc.* **60,** 3061–3070.

Scott, W. J. (1953). Water relations of *Staphylococcus aureus* at 30°C. *Aust. J. Biol. Sci.* **6,** 549–564.

Scott, W. J. (1957). Water relations of food spoilage microorganisms. *Adv. Food Res.* **7,** 83–127.

Scott, W. J. (1962). Available water and microbial growth. *Proc. Low Temp. Microbiol. Symp., 1961,* pp. 89–105.

Wolf, A. V. (1966). "Aqueous Solutions and Body Fluids." Harper (Hoeber), New York.

Wolf, M., Walker, J. E., and Kapsalis, J. G. (1972). Water vapor sorption hysteresis in dehydrated food. *J. Agric. Food Chem.* **20,** 1073–1077.

Wolf, W., Spiess, W. E. L., and Jung, G. (1973). Die Wasserdampfsorptionisothermen einiger in der Literatur bislang wening berücksichtigen Lebensmittel. *Lebensm.-Wiss. u. Technol.* **6,** 94–96.

2

Methods

Until recently, most methods for the measurement of water activity or equilibrium relative humidity (E.R.H.) of foods were basically adaptations of techniques originally designed by or for meteorologists to measure atmospheric humidity. Since nearly all of the procedures used by food scientists to quantify humidity require its measurement in an enclosed atmosphere at equilibrium with a food sample, the extrapolation is essentially one of scale rather than technique.

Interest has increased throughout the food science field on the various aspects of a_w effects in foods, as the implications of a_w control during food processing, packaging, and storage have become apparent. A number of instruments, many produced commercially but all specifically designed to measure a_w levels in foods, have entered the food scientists' instrumental armamentarium. None of these instruments is satisfactory for every conceivable a_w-related application; however, many will produce satisfactory results.

The moisture measurement techniques discussed in this chapter have been classified into three groups based on function: (1) a_w, (2) atmospheric relative humidity, and (3) total moisture. Only the first group actually deals with direct a_w/E.R.H. measurements as related to foods. The second category, ambient relative humidity measurements, covers food-related activities, such as the measurement of relative humidity in food storage areas. These methods are not intended or used primarily for the quantification of water activity. The third group, total moisture measurements, concerns those methods which measure the total amount of water in foods without regard for the "condition" or degree of water binding. These methods relate to water activity through sorption isotherms, and it is for this reason that they are included in this chapter.

None of the categories is intended to be exhaustive. This is especially true with those procedures which measure atmospheric relative humidity and total water content. As new or modified techniques appear in the literature, the compilation presented here may become obsolete. For this reason, the principles on which the various techniques are based (especially a_w procedures) are emphasized, in order to provide a broader, more useful coverage. Two brief sections devoted to method calibration procedures, and a_w control techniques complete this chapter.

There exist a number of excellent reviews that deal with the general subject of humidity and moisture measurements (Wexler, 1965; Wexler and Brombacher, 1951) and with food-related a_w methods (Gough, 1974; Toledo, 1973; Smith, 1971; Szulmayer, 1969) specifically. These sources provide basic information on a variety of methods and are recommended as supplementary reading to those interested in selecting or evaluating available techniques for specific applications.

DESIRED CHARACTERISTICS

The attributes that methods and equipment for a_w measurement should possess do not differ appreciably from those of other analytical procedures. They are (1) accuracy, (2) reproducibility, (3) speed, (4) low cost, (5) portability (specific uses only), (6) ease of use, and (7) durability.

Accuracy and reproducibility are probably of greatest importance, since little value can be given to a procedure which does not determine, with reliability and precision, the experimental variable to be measured. Most methods for a_w estimation or range finding should be accurate to within 0.02 a_w units. Research applications or methods in which it is desirable to predict the storage stability of a food in a critical a_w range may demand much greater accuracy. In such determinations, an accuracy of $\pm$ 0.005 a_w units is often required and attainable by some methods. The analyst should be cautioned that many methods reporting a_w measurements to the third decimal place may be extrapolations of values derived by procedures accurate to only two decimal places. Accuracy greater than $\pm$ 0.005 a_w is probably unnecessary for most food-related applications.

The attribute of portability may be the most tenuous of those listed. In many cases, such as "in-line" a_w measurements in a food process, portability is of little consequence to the food scientist or engineer, whereas durability and ease of sensor cleaning may be far more important. Also, the desired configuration of the instrument may be important. For example, with electric hygrometers, automatic or manually operated switching devices and/or recorders may be required for a particular application and may limit severely the portability of the instrument. Alternatively, ease of handling may be extremely important when "spot checking" humidity in a food storage warehouse, in which case a sling psychrometer or similar device may be most appropriate. In most a_w methods, the foods to be measured must be in equilibrium with the surrounding atmosphere. Usually, it is advisable to keep this space as small as possible to ensure rapid equilibration and a more rapid analysis. Thus, the sensor chamber is ideally kept to minimal dimensions. Again, this factor will vary from method to method.

WATER ACTIVITY METHODS

Graphic Interpolation

One definition of a_w states that this parameter is the equilibrium relative humidity or a_w at which a substance neither gains nor loses moisture at a specific temperature. In practice, this concept may be used to estimate, with reasonable precision, the water activity of an unknown material. Instead of attempting to determine the point of zero water loss or gain, the weight of sample water loss or gain in different relative humidity chambers is measured for a given period of time, usually 1 or 2 hours. If the amounts of water gained or lost by the several aliquots of the sample maintained in different relative humidities are then plotted against a_w, this plot will intersect with the line representing zero moisture change. It is this point of interpolation that represents the a_w of the sample.

Saturated or unsaturated salt solutions may be used to obtain standards of prescribed a_w levels (Table 2.1). Usually solutions differing by very small increments of a_w are used to assure that the amount of water transferred is small in relation to the amount of controlling solution present. The need for several equilibration solutions requires that at least four separate chambers be used in this determination. This may impose space limitations on the average food laboratory.

Landrock and Proctor (1951), who first proposed this method, tested a number

TABLE 2.1.

Water Activity Values (25°C) of Saturated Salt Solutions as Stated in Various Reports from the Literature[a]

Saturated salt solutions	Washburn, 1926	Wexler and Hasegawa, 1954	Rockland, 1960	Weast, 1972–1973	Stokes and Robinson, 1949
$LiCl \cdot H_2O$	0.15	0.124	0.12	0.15	0.11
$KC_2H_3O_2$	0.20	—	0.23	0.20	0.22
$MGCl_2 \cdot 6H_2O$	0.33	0.336	0.33	—	0.33
K_2CO_3	—	—	0.44	0.44	—
$Mg(NO_3)_2 \cdot 6H_2O$	0.55	0.549	0.52	0.52	0.53
NaCl	0.76	0.755	0.75	—	0.75
$(NH_4)_2SO_4$	0.81	0.806	0.79	0.81	—
$CdCl_2$	—	—	0.82	—	—
Li_2SO_4	—	—	0.85	—	—
K_2CrO_4	0.88	—	0.88	0.88	—
KNO_3	0.94	0.932	0.94	—	0.93
Na_2HPO_4	0.95	—	0.98	0.95	—
K_2SO_4	0.97	0.969	0.97	—	—

[a] Labuza *et al.*, 1976.

of food products and found it to be a practical means of a_w measurement. Others (d'Alton, 1969; Smith, 1965) have applied this graphic method to confectionery products, corn starch, and fondant icings with good results.

Bithermal Equilibration

Where a_w data on reference solutions are suspect or not available, the bithermal technique of Stokes (1947) will provide an absolute standard. The vapor pressures of solutions are determined by a method depending on the vapor phase equilibration of the solution at 25°C with pure water at some lower temperature. The apparatus consists of two copper containers ("bells"), each with a flat base, joined by a thin-walled copper tube which permits each bell to be submerged in a separate thermostated water bath. The apparatus is rocked. The bell held at 25°C contains, in a flat-bottomed silver dish, the test solution. The other bell, containing a dish with pure water, is thermostated at a lower temperature. A 100-junction, copper-constantan thermocouple is built into each water bath. When equilibration is reached, the temperature difference between the baths and, hence, between the samples is calculated with considerable precision from the thermocouple emf (E) by the formula:

$$\Delta T = 0.2350\,E + 0.0000690\,E^2$$

The concentration of the equilibrium solution is determined by drying or by analysis. The vapor pressures of water at 25°C and at 25°–ΔT° are read from the International Critical Tables (Washburn, 1926), and the a_w at equilibrium is the ratio of the vapor pressures at the lower and higher temperatures.

Manometry

As noted earlier, the moisture content of a food is related directly to its vapor pressure at a constant temperature. This pressure may be measured accurately by manometric procedures, as was demonstrated by Makower and Myers (1943). Relatively high precision ($\pm$ 0.002 a_w) has been claimed for this technique by Sloan and Labuza (1975a) and Labuza *et al.* (1976).

The sample to be measured is ground and introduced into a flask, which is attached through a trap to a simple manometer. The manometer assembly (Fig. 2.1) is then evacuated, and during this time the sample chamber is maintained at approximately −80°C. Following evacuation, the sample is warmed to room temperature, while one side of the manometer is maintained at essentially zero pressure. The fluid level in the sample arm of the manometer is then deflected by the increase in pressure caused by the vapor pressure of the sample itself. Air entrapped in the sample also contributes to deflection of the manometer, but this can be compensated for relatively easily. Multiple equilibrium relative humidity

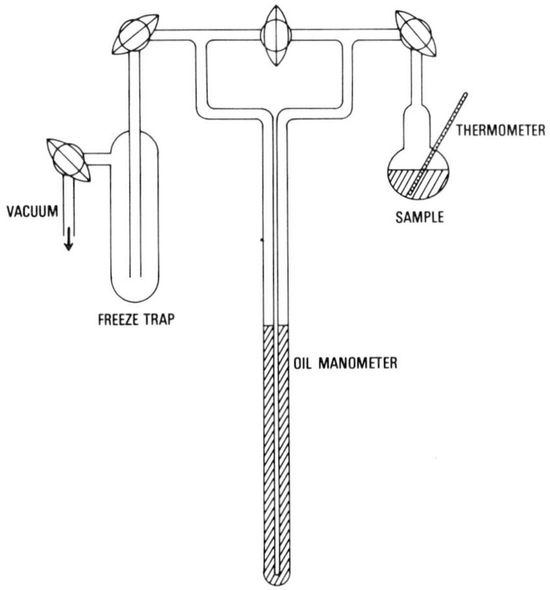

Fig. 2.1. Vapor pressure manometer.

determinations can be carried out on the same sample at various moisture levels by allowing sample moisture to distill into the freeze trap and by reweighing the sample, a modification introduced by Taylor (1961). This author also noted that, although the vapor pressure of fats does not contribute significantly to measurements, equilibration of these substances proceeds slowly. He suggested that such fatty materials be extracted with an appropriate solvent before introduction into the sample chamber. Sood and Heldman (1974) investigated several factors that might influence a_w determinations by the vapor pressure manometer method and noted that sample size, particle size, and sample container volume did not influence the a_w values obtained. These authors reported good precision when 10 replicates, each of a known a_w nonfat dry milk and a similar sample of unknown a_w level, were measured.

Karel and Nickerson (1964) further modified the manometric apparatus to permit the simultaneous measurement of sorption isotherms by addition of a sensitive quartz spring, permitting weight changes to be measured during vapor pressure determinations. Loncin (1955) has used similar manometric devices to determine the a_w measurements and isotherms of oils and fats.

In addition to good precision, manometers are relatively inexpensive. Unfortunately, instruments capable of measuring wide vapor pressure differences are cumbersome and extremely fragile. Also, absolutely uniform temperatures and a high degree of thermometric accuracy are required. This instrument holds considerable promise for routine a_w analyses if these shortcomings can be overcome.

Hair Hygrometry

This procedure relies on the hygroscopicity of human hair and the ability of this material to stretch when hydrated. Hair, usually three or more strands braided, is fixed at one end and attached to a sensitive lever arm at the other. The lever arm is connected to a recorder pen or a dial that reads directly in percent relative humidity. A canister-type hair hygrometer commonly used for a_w determinations in the food industry is shown in Fig. 2.2. These instruments are relatively less sensitive than many a_w instruments ($\pm$ 0.03 a_w) and so are often used in preliminary, range-finding determinations or in applications where their lack of reliability and insensitivity can be tolerated. Additional disadvantages include delay in reaching equilibrium and a tendency to exhibit hysteresis. Their principal attribute is comparatively low cost.

Isopiestic Equilibration

This method, described by Robinson and Sinclair (1934), provides an accurate means of determining the relationship between water sorption isotherms. A test sample and a reference solution, usually using a salt, are equilibrated under vacuum, and the water contents of both are then determined. The a_w of the equilibrated reference system can then be found from standard tables (e.g., as in Robinson and Stokes, 1959), which is thus the a_w of the test material at its final

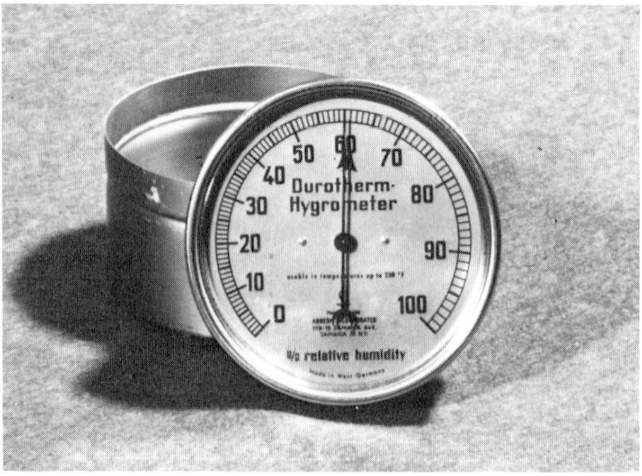

Fig. 2.2. Hair-type mechanical hygrometer.

water content. A series of such determinations with reference solutions having a suitable range of initial concentrations permits the construction of the isotherm. Alternatively, the reference solutions may be saturated solutions containing excess solid phase in which the a_w remains constant, provided that the temperature is constant and that the solution does not absorb sufficient moisture to become unsaturated. In this case, the final water content of the reference solution is not determined.

To minimize temperature gradients between samples and solutions, both are placed in silver dishes in good thermal contact with a copper block. Equilibration proceeds in an evacuated desiccator, which is rocked in an accurately thermostated water bath.

Related procedures employing isopiestic equilibration have recently appeared in the literature (Fett, 1973; Vos and Labuza, 1974). These techniques involve the preparation of a standard sorption isotherm by careful weight-difference determinations of a reference material previously equilibrated with solutions of known a_w. Proteins (casein or soy isolates) (Fett, 1973) and microcrystalline cellulose (Vos and Labuza, 1974) have proved suitable as reference materials.

Following establishment of the standard isotherm, food samples are equilibrated with the reference material, usually in a vacuum desiccator, for 24–28 hours (depending on the a_w range). This material is then carefully weighed to determine its moisture content, and a_w is determined by reference to the standard curve. Comparisons between a_w levels of various foods determined by these techniques and by electric hygrometric procedures have given good correlations at a_w levels > 0.90, and superior precision has been claimed at a_w levels $\leqslant 0.90$.

The principal advantages of these gravimetric techniques are the very low cost of the apparatus, apparent accuracy at high a_w levels, and freedom from contamination. Balanced against this is the need to use several desiccators to obtain replicate determinations, the long periods (24 hours) required for equilibration, and the need to establish a new standard isotherm whenever a different lot of reference material is used.

In principle, this technique resembles the graphic interpolation method of Landrock and Proctor (1951) described earlier in this chapter. The major difference is that in this method, the sample is allowed to come to complete equilibrium with the reference material. Another modification is that of Gur-Arieh *et al.* (1965), in which an air stream, at a given relative humidity, is directed through the sample to attain a more rapid equilibrium.

Electric Hygrometry

Two basic types of electric hygrometers are used in food-related applications. The first is based on the measurement of conductivity or resistance of an hygroscopic salt in equilibrium with an ambient atmosphere. As water is absorbed or desorbed by the salt, its ability to carry current (conductance) is measurably altered. These instruments are referred to as either Gregory or Dunmore hygrometers, depending on the design of the humidity-sensing element.

The second type has been referred to as an electrolytic hygrometer. The operation of this instrument requires that an alternating current be passed through a saturated LiCl solution suspended onto an inert carrier such as glass wool. A current potential difference of 25 V, which heats the cell, is applied across the solution. The water vapor pressure (w.v.p.) of the solution rises, and upon reaching the w.v.p. of the environment, water evaporation occurs. The dried LiCl residue remaining after evaporation no longer conducts current, and heating ceases. As the residue cools, water is once again taken up from the environment, and the cycle is repeated at a reduced amplitude. Eventually, a temperature is reached at

Fig. 2.3. An electric hygrometer. (Sina Instruments, Zurich, Switzerland).

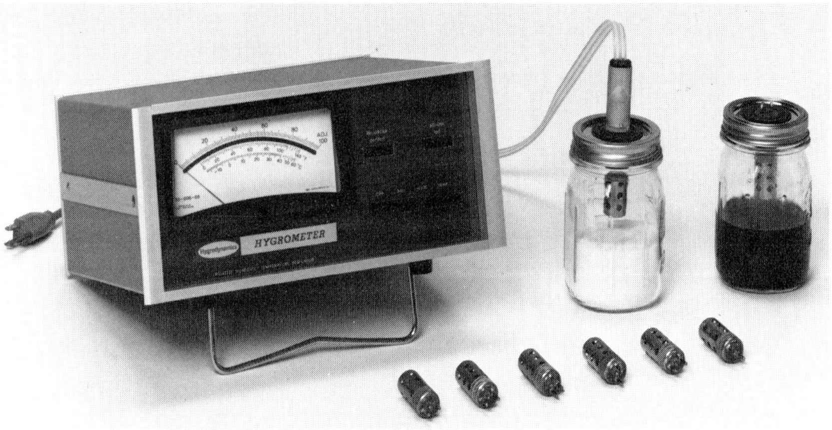

Fig. 2.4. An electric hygrometer (American Instrument Co., Silver Spring, MD).

which the w.v.p. of the solution is equal to the w.v.p. of the environment. This temperature is then measured and related to the w.v.p. of the saturated LiCl solution and hence, the environment. From this, the equilibrium relative humidity of the environment can be calculated.

Mossel and van Kuijk (1955) used an instrument of this type to measure the equilibrium relative humidity of a number of foods, sodium chloride solutions of known composition, and saturated solutions of several salts. In all cases, observed relative humidities correlated well with established values for these materials. In addition, these authors stressed the rapidity of determinations as compared to other methods.

The other type of electric hygrometer, mentioned earlier in this section, requires the measurement of electrical conductivity or capacitance of a hygroscopic substance. Usually this substance is a salt, such as LiCl. However, the anodized surface of an aluminum rod has also been used.

The former of these two types of instruments, requiring the measurement of conductance and resistance across a hydrated material, such as LiCl, is probably the instrument used most frequently to measure the water activity of foods. Figures 2.3 and 2.4 illustrate two commercial instruments of this type. The instrument in Fig. 2.3 employs an integral heating circuit that allows the direct reading of a_w measurements from the dial indicator. The instrument in Fig. 2.4 requires that indicated readings be referred to a graph of dial readings versus water activities for a number of temperatures.

In Dunmore-type electric hygrometers, a food sample is brought into equilibrium with an enclosed volume of gas. The sensor, a hygroscopic salt conducting a small electric current, is also exposed to the gas. The amount of current passing through the salt solution or the resistance in the solution is a function of the degree of salt hydration and is measured.

Sensors of this type have been constructed in a variety of configurations, some of which allow measurements at multiple remote locations through automatic switching circuits, in-line measurement of free-flowing materials or special adapters for the a_w measurement of packaging materials. In addition, Dunmore sensors have been described (Brastad and Borchardt, 1953) which can be fitted into the end of a hypodermic needle to permit a_w measurements of the interior of individual grains of wheat or other very minute samples.

The principal disadvantages of these instruments are their high cost and the tendency of the hygroscopic salts to become contaminated with polar materials such as glycols. An example of the latter is shown in Fig. 2.5. Usually, contamination from glycols results in a reduced a_w, although erratic readings may occur.

Aging effects resulting from physical and chemical changes in the electrode or

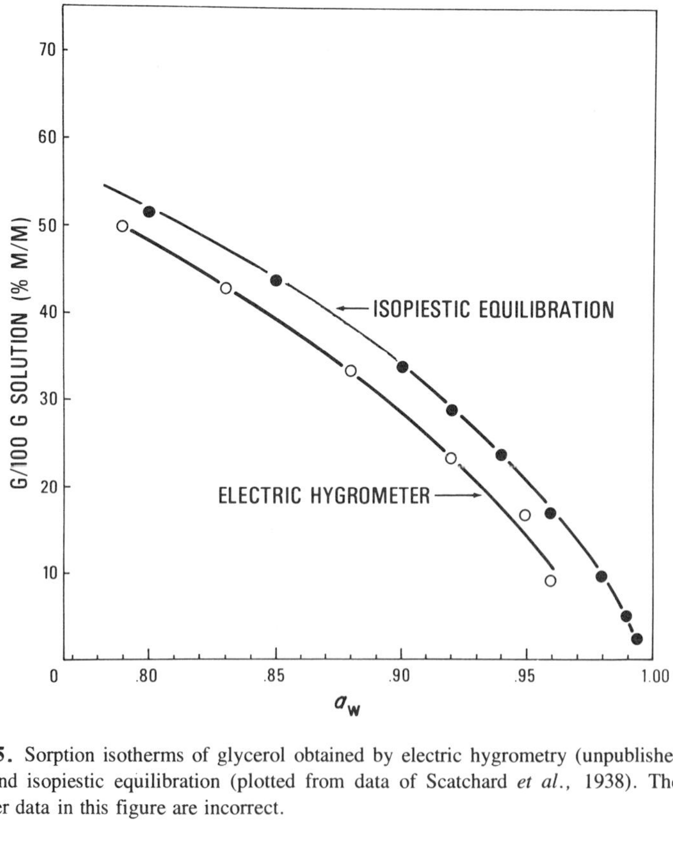

Fig. 2.5. Sorption isotherms of glycerol obtained by electric hygrometry (unpublished data of Troller) and isopiestic equilibration (plotted from data of Scatchard *et al.*, 1938). The electric hygrometer data in this figure are incorrect.

the moisture-sensitive film (Kobayashi and Toyama, 1965) can also result in errors. Finally, water droplets (from splashing or condensation) will adversely affect sensor response. Usually, this problem can be avoided by proper sensor design and/or by providing some means of equalizing the temperatures of the sample and sensor.

Filters that absorb contaminants are available, or can be manufactured, but these extend the time required for sensor equlibration. Usually, "resting" the sensors for several days at low relative humidities (25–30%), followed by precise recalibration, restore accuracy after contamination or sensor "fatigue."

The equilibration rate of the sample chamber is difficult to predict and depends on a number of factors, such as the volume of the enclosed space between sample and sensor, the nature of the sample, and the a_w range. Usually, equilibration occurs most rapidly at a_w levels < 0.90. Electric hygrometers measure conductance of an aqueous electrolyte solution in terms of a current flow caused by an applied alternating current potential. This current can be amplified and recorded if desired. Recorders may be integral with or connected to the hygrometer and provide a convenient and accurate means of determining the point at which equilibration occurs. Most equlibration periods, whether determined by a recorder or point-by-point determinations, require 15 minutes to 2 hours.

Using a Dunmore instrument of the type illustrated in Fig. 2.3, Karan-Djurdjic and Leistner (1970) measured the a_w of several saturated salt solutions within a range of 0.75–0.92 a_w. Measurements obtained using two identical instruments did not agree; however, the finding of approximately constant error factors when compared to the "known," standard values allowed application of a correction factor. A comparison of the Dunmore-type instrument with the graphic interpolation method of Landrock and Proctor (1951) described earlier in this chapter revealed an average deviation between the two methods of 0.0023 a_w for several types of sausages. Although the precision or reproducibility of measurements obtained by interpolation was slightly better than those determined by the hygrometer (after correction), these authors believed that the precision obtained was satisfactory. A statistical evaluation of an electric hygrometer was reported by Troller (1977), who noted that this instrument produced acceptable accuracy and excellent precision throughout the a_w range studied (0.75–0.97), providing that the instrument was carefully calibrated. This author did not attempt to compare the hygrometer with other a_w-measuring devices (see reference to Labuza *et al.*, 1976, elsewhere in this chapter). Sensor-to-sensor variations were negligible, and excellent precision was maintained throughout the a_w range studied if duplicate determinations were made.

The relatively good precision, ease of operation, and rapid nature of hygrometric determinations usually outweigh the drawbacks associated with hysteresis and have resulted in the widespread application of these instruments in industry and in laboratories. In many cases, in-line or in-process determinations are possible,

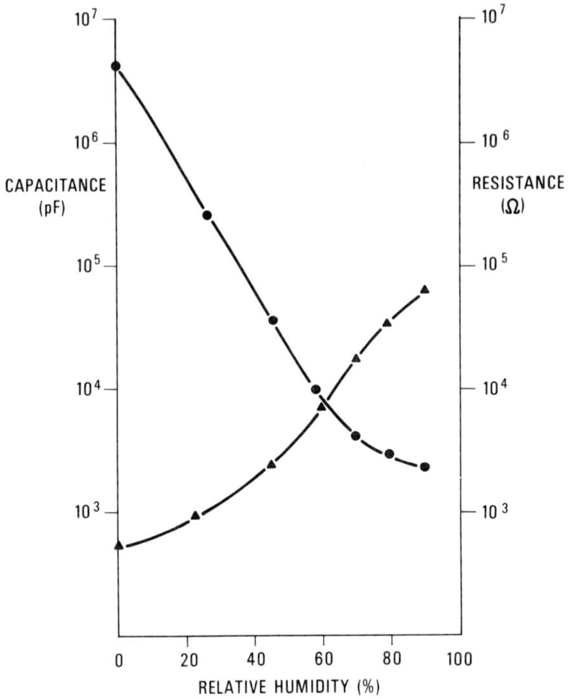

Fig. 2.6. Variation in capacitance (●) and resistance (▲) with relative humidity (Cutting *et al.*, 1955).

and sensors of diverse geometry have been developed for a variety of applications. As with all indirect methods, however, hygrometers must be carefully calibrated to obtain satisfactory results.

The development of capacitance–resistance hygrometers is more recent than that of Dunmore-type hygrometers and has come about largely through the work of A. C. Jason and his colleagues at the Torry Research Station in Scotland. Instead of absorbing atmospheric water in a hygroscopic salt, the sensor of this instrument utilizes a porous film of Al_2O_3 (formed by acidic anodization) to adsorb water. This film is "sandwiched" between an outer conductor (usually graphite) and metal and acts as an electrical transducer or dielectric, which responds to changes in relative humidity by changes in capacitance (Fig. 2.6). Cutting *et al.* (1955) have shown that the response of sensors of this type is relatively independent of temperature in the range 0°–80°C and is amenable to control and/or recording functions. These sensors also have the advantage that they exhibit negligible hysteresis. Accuracy is about ± 0.02 a_w units, a level which is somewhat lower than that for Dunmore-type hygrometers. However, recent improvements, such as replacement of the colloidal graphite layer with aluminum (Jason, 1965) or gold (Miyata and Watari, 1965) and implementation of the Al_2O_3 layer with sodium tungstate, have improved both accuracy and long-term stability of these sensors. Several types of these instruments are available commercially.

Chemical Methods

Certain chemical compounds change in color as they absorb water from an atmosphere of high relative humidity. For convenient handling and observation, these compounds (e.g., cobaltous bromide, chloride, or thiocyanate) may be impregnated in a strip of absorbent paper. The activated or hydrated strip changes color in relation to the amount of imbibed moisture and is compared to a color reference chart, much as pH-sensitive paper is used.

Originally, these systems were used in biological fields to measure transpiration rates on the surfaces of leaves; however, later applications in foods-related areas have arisen (Pomeranz and Lindner, 1953; Ward and Tischer, 1953). Although the lack of accuracy (to ± 15.8% at low relative humidity) and long equilibration periods (Solomon, 1957) limit the use of these methods to applications in which only rough approximations are needed, the inexpensive nature of the test and its ease of use recommend it for certain applications.

Freezing Point Depression

The determination of a_w levels of solutions can also be accomplished by the accurate determination of depression in freezing point of these solutions relative to pure water. The vapor pressures of ice at various temperatures are published in chemical handbooks and can be related to the vapor pressure of supercooled water. Wodzinski and Frazier (1960) utilized this technique to characterize some of the factors which interact with a_w to suppress growth of *Pseudomonas fluorescens*.

The exact temperature at which ice crystals begin to form in a solution, e.g., bacteriological media, is often difficult to determine, and special equipment utilizing highly accurate thermometers has been developed for this purpose (Fig. 2.7). Ayerst (1965a) stated that one of the principal disadvantages of this method is the difficulty of predicting a_w levels at normal microbiological growth temperatures from data obtained at temperatures near 0°C. In addition, this method is applicable only to solutions and is not useful for measuring the a_w of solid foods.

Similar restrictions apply to methods employing boiling point elevation determinations. Theoretically, the boiling point of an ideal solution increases (vapor pressure at a given temperature decreases) in relation to the quantity of solute present in the system. These changes can be related to data in commonly avail-

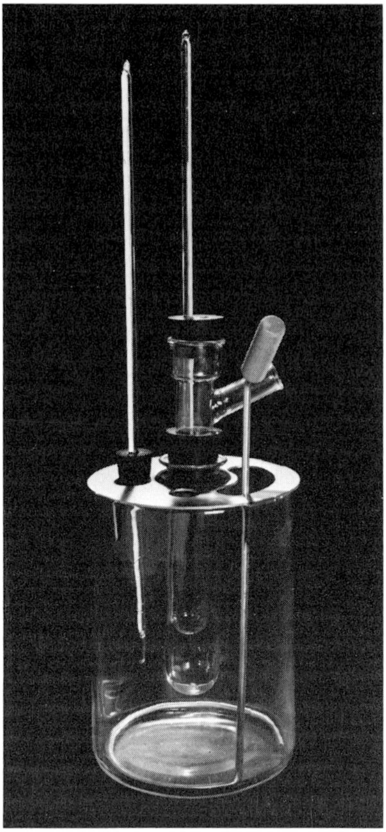

Fig. 2.7. Freezing-point apparatus.

able handbooks and water activity calculations. Insofar as the present authors are aware, freezing/boiling point methods are not applied widely in the food industry.

Dew Point Methods

Another means of measuring vapor pressure is to determine the exact temperature at which condensation of water vapor occurs. This parameter is defined as the dew point. The basic principle involved in dew point determinations is that air may be cooled without change in water content until saturation is reached. The temperature at which this saturation is achieved can be determined by observing condensation on a smooth, cooled surface such as a mirror or sight glass. This dew point temperature is related to vapor pressure, relative humidity, and

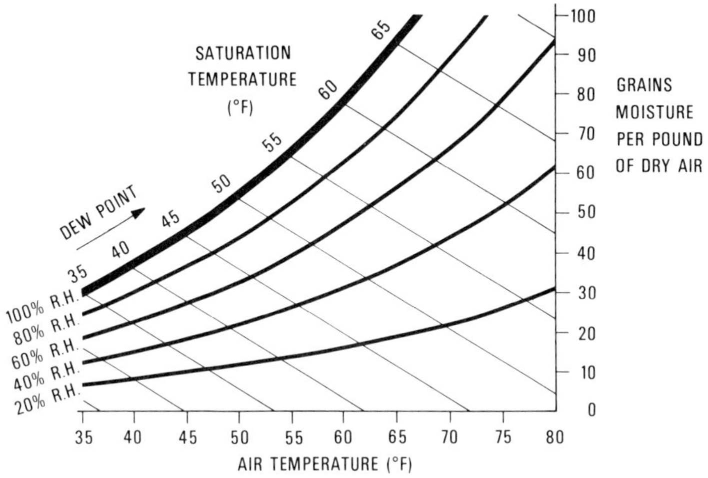

Fig. 2.8. Psychrometric chart.

water activity by reference to psychrometric charts, an example of which is shown in Fig. 2.8. Numerous instruments are available for determining dew point temperatures. Various refinements of these devices have been developed, most of which relate to obtaining the exact point at which condensation first appears. In contemporary automatic devices (see diagram Fig. 2.9), a gas (of unknown vapor pressure and in equilibrium with the sample to be measured) is exposed to a mirror cooled in some manner, usually by a thermoelectric (Peltier effect) cooler. A beam of light is directed onto this mirror and reflected into a photodetector cell. When condensation occurs on the mirror, the photodetector senses the resulting change in reflectance and decreases the current to the thermoelectric cooler, eventually resulting in the establishment of an equilibrium. The mirror

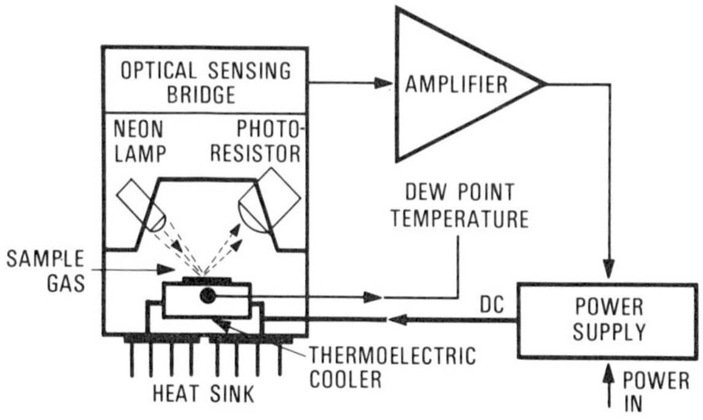

Fig. 2.9. Diagram of an automated dew point device (courtesy EG & G Co., Waltham, Mass.).

temperature at which this equilibrium is obtained can be accurately determined and related to the relative humidity or vapor pressure of the unknown gas. Although these instruments are ideally suited to gas streams, static atmospheres in equilibrium with food samples can also be evaluated. However, one must assure that the volume of water condensed on the mirror will not cause a significant depression in the relative humidity of the air in equilibrium with the food sample.

Rodel and Leistner (1972) developed and tested a humidity-sensing chamber that is adaptable to an instrument of this type. With this system, the a_w of small quantities of saturated salt solutions of known a_w, and, also, meats and meat products, were measured with considerable precision. Water activity measurements of different sausages revealed close agreement with those obtained using a commercially available electric hygrometer.

Commercial dew point instruments usually obtain nominal dew point temperature accuracy of $\pm$ 0.4°F, depending upon the instrument and the care with which they are operated. Ayerst (1965b) has stated that with proper precautions $\pm$ 0.01 a_w is easily obtained and $\pm$ 0.005 is attainable. It is extremely critical that mirror or viewing surfaces be clean and uncontaminated.

Less sophisticated dew point instruments, which rely on visual observation of condensation on the mirror surface, have been described. These instruments may (Anagnostopoulos, 1973) or may not (Ayerst, 1965a) rely on thermoelectric, Peltier-effect cooling modules to reduce the temperature of the condensing surface.

Bulk Effect Instruments

The principal bulk effect instrument currently in use to measure a_w is the Brady array unit. The sensor of this instrument consists of a series of precisely arrayed crystal lattice structures which act as semiconductors. Water molecules cause a stress to occur in the bonds in the crystalline lattice. As the bonds become disoriented, energy is released, producing a change in conductivity. This change is sensed by a deviation in signal applied to the sensor by an oscillator. The extent of signal alteration is related to the amount of water in the lattice and is detected by a voltmeter after demodulation. Usually, the food to be analyzed is placed in a sealed chamber containing the sensor, so that equilibrium conditions can be established between the lattice and the food.

The performance of this instrument is described in the publications of Little *et al.* (1974) and Labuza *et al.* (1976). In the latter report, the Brady array instrument, when compared with other a_w-measuring devices, showed the greatest average difference from theoretical a_w values of saturated salt solutions (a_w range from 0.33 to 0.97) among the various instruments tested.

Some investigators have substituted thermocouples for thermometers, usually

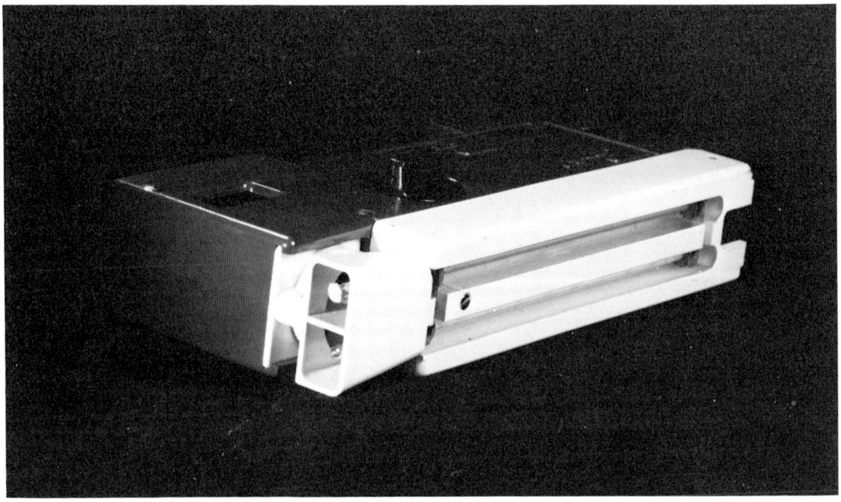

Fig. 2.10. Assman-type hygrometer.

wetting one thermocouple with either a drop of water or a moistened wick. The principles involved are exactly the same. These devices are relatively expensive, but also are very compact and may be useful in measuring relative humidity under conditions in which thermometers cannot be conveniently used.

RELATIVE HUMIDITY METHODS

Thermometric Methods

The water vapor pressure or water-holding capacity of a gas may be obtained by measuring its ability to absorb additional water. In the wet bulb–dry bulb

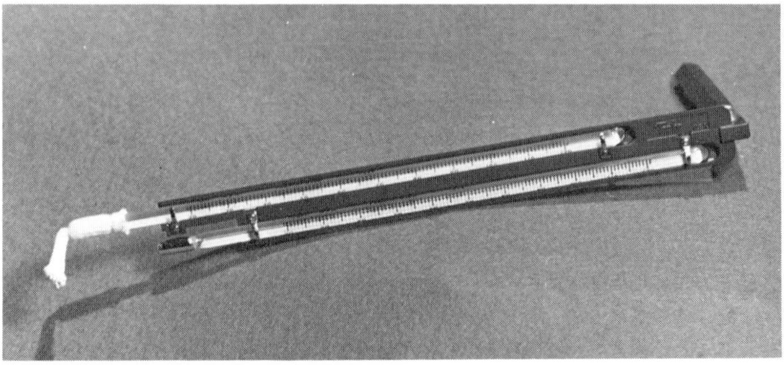

Fig. 2.11. Sling psychrometer.

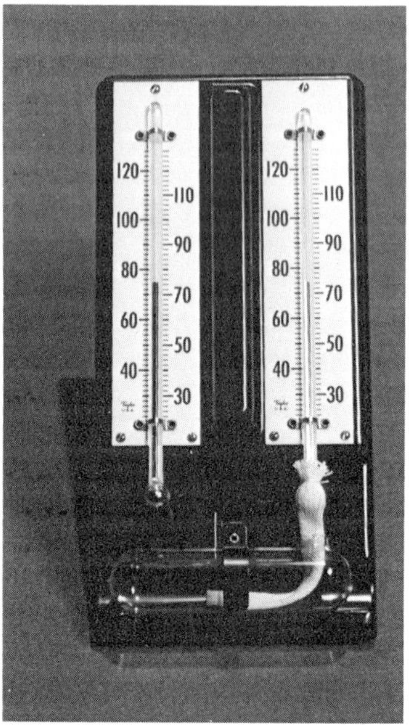

Fig. 2.12. Wall-mounted wet/dry bulb hygrometer (Mason's form).

thermometric technique, the temperatures indicated by a matched pair of thermometers, the bulb of one exposed to ambient conditions and the bulb of the other moistened, are compared. The rate of evaporation of water from the wet bulb is dependent on the vapor pressure of the gas surrounding it. This evaporation cools the thermometer bulb. If the air is relatively dry, evaporation is rapid, and the wet bulb is cooled to a greater degree. The wet and dry bulb temperatures thus obtained are related to vapor pressure: vapor pressure $= P_w - AP(t - t_w)$; $t =$ ambient temperature; $t_w =$ wet bulb temperature; $P =$ barometric pressure; $P_w =$ water vapor pressure on saturation at the wet bulb temperature; $A =$ gas flow rate factor (wet bulb).

The calculations in the above equation can be avoided by referring to tables which relate relative humidity to wet and dry bulb temperatures or to the temperature depression at the wet bulb. In practice, the above equation is independent of gas flow rate (factor A) if this rate exceeds 3 meters/second. Thus, provision in wet/dry bulb methods is often made for movement of the gas past the wet bulb, as in the Assman-type hygrometer, in which a small, battery-driven fan circulates air past the wet bulb (Fig. 2.10). Alternatively, the thermometers may be moved through the air. A commonly used instrument of the latter type is the sling psychrometer shown in Fig. 2.11. A typical wall-hung hygrometer, which does not provide for gas circulation, is shown in Fig. 2.12. While not as accurate as either the Assman or sling hygrometer, this device may be adequate for the measurement of ambient relative humidity in food storage areas.

TOTAL MOISTURE METHODS

Gravimetric

There are two basic methods involving gravimetry by which information relating to the moisture content and/or condition of a material can be obtained. The first is simply to dry a preweighed sample to a constant weight under carefully controlled conditions. Assuming that only water leaves the sample, percent moisture (weight/weight) can be calculated. Although simple in concept and execution, moisture estimations obtained gravimetrically are subject to error from many sources (Oxley and Pixton, 1961). One common error results from the formation of a crust or "case-hardening" of the sample particles, a process which prevents complete escape of water vapor from their interior. The extent of sample grinding and composition are important factors also.

Foods are normally dried for specified periods, but under some conditions errors may be introduced as a result of incomplete water removal. It is probably preferable in many cases to dry to a constant weight, although this is a procedure which may require relatively long heating periods. In such instances, care must be taken to assure that sample decomposition does not influence results. Furthermore, constant weight may be unattainable because water may be slowly produced by Maillard reactions. To hasten complete drying of samples, vacuum ovens may be used. In most methods of this type, drying to constant weight is practiced.

Another type of gravimetric technique (Bosin and Easthouse, 1970) employs an equilibration chamber in which a sample is brought to a given moisture level by equilibration with a saturated salt solution of the desired relative humidity. While enclosed in the chamber, the sample may be weighed by an overhead balance. Equilibration is accomplished rapidly (about 3–15 times faster than in static systems) with the aid of an interior fan driven by an external magnetic stirrer. Although not specifically stated in the above article, the chamber should probably be elevated slightly above the magnetic stirrer to provide an insulating air gap and therefore minimize errors due to heating.

Gas Chromatography

Gas chromatography has been applied to the moisture analysis of grain by a number of workers (Weise *et al.* 1965; Schwecke and Nelson, 1964). Samples to be analyzed are usually ground or blended in the presence of absolute methanol

TABLE 2.2.

Moisture Contents of Fruits and Fruit Products as Determined by Vacuum Oven and Gas–Liquid Chromatography[a]

	Average moisture (%)[b]	
Product	Vacuum oven	Gas chromatography
Guava	87.30	87.01
Papaya	87.19	86.60
Fresh banana	69.66	70.33
Air-dried banana	18.37	18.93
Drum-dried banana	4.23	4.13
Freeze-dried banana	4.29	4.51
Raisin	15.46	16.94

[a] Brekke and Conrad, 1965.
[b] Quadruplicate assays.

or *sec*-butanol plus methanol. In the former case, the peak area measurement is obtained and related to a series of standards containing known amounts of water in methanol. In the latter case, the ratio of the peak height of water to the measured peak height of *sec*-butanol is obtained and converted to the weight ratio of water to *sec*-butanol using a calibration curve derived from known solutions. The weight ratio is multiplied by the volume of *sec*-butanol and percent moisture obtained by conversion.

A similar method, employing *sec*-butanol as an internal standard and methanol as an extractant, was used by Brekke and Conrad (1965) to determine the moisture content of fruits and fruit products. The results obtained agreed very well (Table 2.2) with a vacuum oven technique. This method appears to meet many of the criteria noted earlier in this chapter and is applicable to cereals and fruits. Like other methods employing an extraction step, precision and sensitivity may be severely limited, depending on the thoroughness of extraction. In addition, the accuracy of the analysis depends on preventing the water peak from "tailing." This is normally accomplished with Teflon powder columns impregnated with various stationary phases.

Karl Fischer Titration

In 1935, Karl Fischer, the German chemist, introduced a reagent specific for water that contained pyridine, methanol, sulfur dioxide, and iodine. When combined with water, the following two-stage reaction (Wernimont and Hopkinson, 1943) occurs:

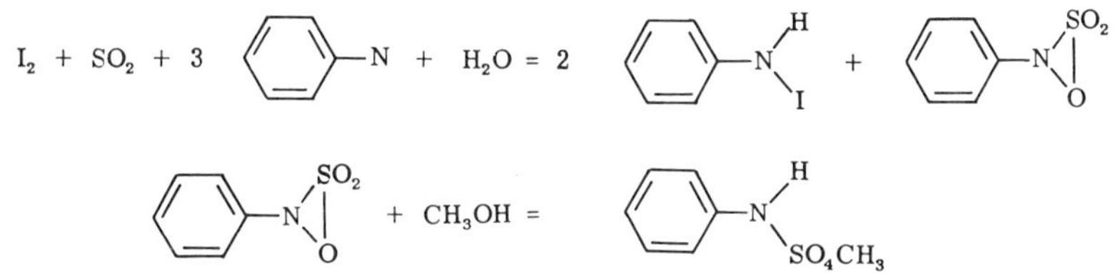

The presence of excess I_2, detected visually or potentiometrically, indicates the endpoint of the reaction. Back titrations with methanol–water have also been used.

Pomeranz and Meloan (1971) have stated that this method is especially useful for the analysis of dried fruits and vegetables, candies, roasted coffee, and fats. Although a relatively old procedure for moisture determination, the Karl Fischer titration remains the recommended method for water analysis of dehydrated vegetables in some official manuals (e.g., A.O.A.C, 1975).

In addition to its remarkable specificity for water, this titration is rapid, requiring from 10 to 60 minutes to complete, a great improvement over oven-drying procedures (4–10 hours) which were the only alternative techniques available in 1935. Interference in the Karl Fischer titration has been observed in the presence of aldehydes and ketones (which react with methanol to release water resulting in a fading endpoint), metal oxides, borates, sulfides, and hydroxides (Fosnot and Haman, 1945). Additionally, the kind of material being analyzed and the particle size of the sample can be very important factors. To improve the efficiency of extraction, many workers have substituted dimethylformamide for methanol.

Although supplanted in many laboratories by more broadly applicable techniques such as infrared (see below), the Karl Fischer titration continues to be widely used 40 years after its discovery.

Nuclear Magnetic Resonance

The application of nuclear magnetic resonance (NMR) techniques to moisture measurements of foods occurred very shortly after the discovery of NMR by Bloch *et al.* (1946) and Purcell *et al.* (1946). Initially, equipment limitations prevented or complicated these techniques. Later, as commercial instruments appeared, possessing broad, quantitative capabilities, interest in this procedure as a routine means of measuring food moisture increased (Shaw *et al.*, 1953; Shaw and Elsken, 1956; Palmer and Elsken, 1956).

The NMR technique is based on the fact that hydrogen atoms in water have nuclei which possess magnetic properties. That is, they behave as small bar magnets, creating their own magnetic fields. If placed in an external magnetic field, the axes of these nuclei tend to become oriented in a specific, fixed direction with regard to the applied field. If these nuclei are exposed to a superimposed or additional oscillating magnetic field of a specific frequency, they become reoriented. The amount of energy absorbed during this process is related to the number of hydrogen nuclei in the magnetic field (and thus the amount of water) and can be measured, relative to a standard, from the amplitude or area of a voltage signal.

The application of NMR techniques has been extended to the determination of bound and free water in foods. Bound water, by one definition, is water that does not freeze. The NMR procedure may be used to distinguish between these two types of water by two methods:

1. *Freezing*. In this method, NMR distinguishes between frozen (free) and unfrozen (bound) water at a temperature < 0°C. The basis of this technique is that water which is bound in a food normally is associated strongly with macromolecules. In this condition, the water is unable to conform to an ice structure (i.e., to freeze) at temperatures below 0°C. On the other hand, unassociated or free water does freeze. NMR distinguishes between ice and water because the signal exhibited by ice is much broader than that of unfrozen water. This difference can be perceived by signal amplitude differences at specific frequencies. Toledo *et al*. (1968) used this technique to determine the bound water content of wheat flour doughs and found that the values obtained agreed closely with published data.

2. *Room Temperature*. Another characteristic of bound water is its specific molecular mobility at room temperature. In other words, it exhibits a broader signal or shorter nuclear spin relaxation time than free water. Again, this difference in signal can be measured. This method requires measurements at only one temperature as opposed to the freezing technique discussed above; however, in many foods, free and bound water exchange rapidly, creating only one signal. The line width of this signal is therefore a weighted average of the free and bound water signals necessitating measurements on samples of several different water contents. A modification of this procedure enabled Sudhakar *et al*. (1970) to determine the bound water content of wheat flour, cornstarch, and egg white at room temperature without prior freezing.

The principal advantages of the NMR technique, in addition to its ability to differentiate between bound and unbound water, are: rapid (1 minute) determinations, accuracy, nondestructive nature and great precision. Disadvantages are the cost of operating and maintaining the NMR instrument and the need to obtain precise calibration curves (Miller and Kaslow, 1963).

Thermal Analysis

Like NMR, thermal analysis techniques find their principal applications in defining the state or binding of water in foods, rather than in the quantification of the amount of water present. The basis for their inclusion in this chapter is the increasing emphasis that thermal analysis is receiving as a practical indicator of storage stability and the energy requirements for bound water removal during drying. Many of these applications were pioneered by R. B. Duckworth and his associates (Duckworth, 1971; Parducci and Duckworth, 1972) at the University of Strathclyde.

There are basically two types of thermal analysis techniques currently employed. These procedures are differential scanning calorimetry (DSC) and differential thermal analysis (DTA). Some confusion has arisen relative to the use of the former term. For the purposes of our brief discussions, the definitions as stated by Wendlandt (1974) are appropriate. This author has described DSC as a technique for recording the energy required to establish a zero temperature difference between a substance and a reference material. Differential thermal analysis, on the other hand, measures the difference in temperature between a substance and a reference or thermally inert material when these two materials are heated or cooled at identical rates.

In both techniques, the occurrence of these energy-active reactions is demonstrated as a series of peaks or irregularities in plots of temperature difference (Δ_t) versus temperature. Applications and descriptions of several thermal analysis techniques are reviewed by Blain (1975).

Moisture Evolution Analysis

This method of quantifying water content is based on the amount of electrical energy required to electrolyze water evolved from a heated food sample. A weighed food sample is dried by heating in an oven for a specified period. The oven is purged with a nitrogen carrier gas that is transported to a phosphorus pentoxide coulometric detector cell. Moisture in the carrier gas is then electrolytically decomposed to hydrogen and oxygen. The current required to complete the electrolysis is proportional to the moisture content, and, in some instruments, this can be read directly as micrograms of water.

This method is relatively rapid (20–60 minutes for most foods) and simple to use. Limitations are the costly nature of the instrument and the interference of evolved ammonia. Current food applications have involved the measurement of moisture content of sugar and cereal products.

Infrared

Water exhibits a number of absorption bands in the near infrared (IR) region of the spectrum. Bands at 1.93, 1.45, and 0.977 mμ have been most widely used by investigators applying this method for water measurement in foods. Direct (Swift, 1971a,b) and indirect (Vornheder and Brabbs, 1970; Vornhof and

Thomas, 1970) measurements are possible by this technique. In the former, the ratio of reflected intensities of the 1.93 and 1.45 mμ bands is most often used as a measure of water content. In the latter (indirect) technique, water is extracted from a food by dimethylsulfoxide, dimethylformamide, or methanol before IR measurements are taken. Other solvents exhibiting a high degree of hygroscopicity and minimal absorbance in the IR range can be used as extracting agents in addition to those mentioned above.

Direct measurements are somewhat hampered by pigments and food particle size and shape, factors which may introduce errors that must be compensated for by calibration. These problems are not as serious with the indirect technique; however, the accuracy of the latter procedure is dependent on the completeness of extraction. Spectrophotometers with infrared capability are expensive and complex instruments, a condition which further limits the extent of their use in routine food applications. On the other hand, the high degree of accuracy and possibilities for in-line process moisture monitoring have aroused considerable interest in the application of this method to food uses.

The relationship of IR moisture reflectance to water activity appears to be somewhat uncertain at the present time. Swift (1971b) has stated that the degree of water binding influences the reflectance of IR radiation and, thus, the "sensitivity" of these measurements. If this is true, perhaps there is some prospect that this change in reflectance can be correlated to water activity. One can also visualize combining the IR and dew point instruments so that moisture, when condensed on an absorbing surface, can be detected electronically by an IR reflectance monitor, rather than visually.

Another application of infrared radiation is in gravimetric moisture analyses, in which broad-band radiation is used to dry foods, which are subsequently weighed to determine percent moisture content.

Vacuum-Oven Drying

Many standard procedures (i.e., A.O.A.C., 1975) for moisture analysis of various foods specify drying in a vacuum oven. A ground food sample is heated at 70°C for 6 hours under controlled atmospheric conditions. Following the drying period, percent moisture is calculated from the weight loss that occurred. Makower and Myers (1943) have pointed out that this method is subject to numerous errors, such as insufficient drying time to allow many foods to come to a steady state; other factors that cause errors are influence of grinding (particle-size distribution), and air flow. Although these authors concluded that vacuum-oven drying does not provide a satisfactory estimate of the true moisture content, it remains a standard method throughout the world for the determination of the total moisture content of many foods.

Solvent Extraction

Limited use has been made of a method (Mallett *et al.*, 1974) in which water from a food sample is extracted with benzene and measured by Karl Fischer titration. The amount of water thus determined is related to a_w by the equation:

$$a_w = \frac{H_2O_n}{H_2O_o}$$

H_2O_n = amount of water extracted from food; H_2O_o = amount of water extracted from pure water. In a_w measurements of pet foods, this procedure produces results in good agreement with an electric, hygrometric technique. However, some doubt remains as to its applicability to low-a_w foods.

CALIBRATION

The preparation of standards by which measurement devices may be calibrated is a relatively easy task. The need for such calibration is common to all of the methods described in this chapter, and frequency of calibration depends on the method and the manner in which the method is used. Obviously, even very frequent calibration of an instrument with inherently poor sensitivity will not improve accuracy to any appreciable degree.

One of the most common means of calibrating a_w instruments is to expose the humidity-sensing portion of the instrument to chemical solutions of known composition. Saturated solutions of inorganic salts are most frequently used for this purpose, primarily because data relating relative humidity (R.H. = $a_w \times 100$) to these solutions have been published (Rockland, 1960; Wexler and Hasegawa, 1954).

Unfortunately, various reports in the literature do not agree on the exact a_w of saturated salt solutions, and so it is difficult to calibrate instruments and to evaluate a_w measuring devices for accuracy (Troller, 1977). Many workers, however, accept the data on saturated solutions compiled by Stokes and Robinson (1949) (see also Table 2.1) as the best currently available.

Labuza *et al.* (1976) published a compilation of the a_w levels of some saturated salt solutions reported by several sources. This compilation, in abbreviated and altered form, is shown in Table 2.1. As noted by these authors, there is a need for the establishment of unequivocal standard solutions. This need is especially acute if trends continue toward the establishment of regulatory standards based on a_w specifications.

In performing instrument calibrations, it is important to ensure that saturation is achieved. This is usually accomplished by the addition of excess salt and

TABLE 2.3.

Relative Humidity Obtained from Water–Sulfuric Acid Solutions[a]

Relative humidity (%)	H_2SO_4 by weight (%) 0°C	25°C	50°C	75°C
10	63.1	64.8	66.6	68.3
25	54.3	55.9	57.5	59.0
35	49.4	50.9	52.5	54.0
50	42.1	43.4	44.8	46.2
65	34.8	36.0	37.1	38.3
75	29.4	30.4	31.4	32.4
90	17.8	18.5	19.2	20.0

[a] Wexler and Brombacher, 1951.

allowing sufficient time at the calibration temperature for complete saturation to occur. The presence of excess crystals in the saturated solution also provides a convenient visual means of assuring the analyst that the solution is indeed saturated. Care should also be taken to ensure that salt crystals do not protrude above the surface of the solution. The temperature should be carefully controlled, and the sample and solution must be at equilibrium, with regard to both temperature and relative humidity in the measured air space. In addition, the calibrating solution should be within the expected range of a_w values.

Solutions containing less than saturated concentrations of a given solute may also be used to calibrate a_w instruments. Normally nonsaturated solutions are less reliable than their saturated counterparts because weighing errors may occur or because, during periods of exposure to the atmosphere, the solution may gain or lose water and thus become unsatisfactory. This is also true of sulfuric acid solutions frequently used to calibrate a_w instruments; however, in this case, the solutions may be checked by titration.

As noted in Table 2.3, temperature strongly influences the relative humidity of sulfuric acid solutions. This also is true of other humectants (Young, 1967), and so it is important to consider temperature in all calibration procedures.

Because of the ease of formulating calibrating solutions, commercial markets for such systems have not been developed. Very often, however, purchased instruments are accompanied by calibration salt "tablets" as accessories.

CONTROL OF a_w

The control of a_w in foods is essential for maintenance of food wholesomeness, safety, texture, for the suppression of undesirable enzymatic and chemical changes, and for a host of other reasons. Normally, the objective is to satisfy the above criteria without extensively changing the food from its natural state, either with or without subsequent hydration. The degree to which this is achieved ultimately determines the feasibility of utilizing a_w adjustment as a legitimate means of food preservation.

Factors Affecting Control

The prime requisite determining suitability of an a_w-adjusting material in a food is its safety and wholesomeness. Both acute and chronic toxicity, as well as a number of additional toxicological requirements, must be met before a given humectant may be considered for official approval for use at the concentration required.

Cost also influences the type of solute which may find application in a specific food. Generally, NaCl is the least expensive of all humectants, both from the standpoint of overall price and cost effectiveness. Other commonly used humectants, such as glycerol or sucrose, vary greatly in price, depending upon market conditions extant at a particular time or place.

Another, although commonly overlooked, factor in selecting a humectant is its effect on food nutritive value. The effect of a_w on nutritional quality, such as vitamin stability, is discussed in another chapter of this volume, but the simple diluting effect of humectants in lowering the nutritive quality of a food should also be considered. Humectants are generally devoid of nutritional attributes (other than calories), so that their addition in the relatively large quantities normally required to achieve a_w limitation may appreciably dilute the vitamin, protein, or mineral content.

Flavor considerations also are important, and here the type of food must be considered. For example, the amount of glycerol required to preserve a given food may change the overall flavor impression of a relatively bland food. Combinations of humectants may be useful, since the flavor impression of each component may be individually perceived, whereas the reduction in a_w is additive. If, on the other hand, the water-binding capacity of the food itself is utilized to suppress a_w (e.g., in drying), flavor intensity may in some cases be increased, whereas extraneous flavor notes caused by humectants will be absent.

The factors influencing the ability of salt to control a_w levels in foods have been reviewed by Young (1967). Impurities are seldom present in salts at concentrations $> 1\%$, and little effect on a_w levels normally is encountered from this source. Stability and corrosiveness, depending on the salts involved, may on occasion be factors.

Many of the above factors that restrict the use of humectants in foods may not apply in experimental systems. For example, an investigator determining the optimal a_w for the activity of α-amylase is concerned primarily with the precise a_w of his test system and the intrinsic effect of the humectants on the enzyme reaction rather than its flavor, costs, etc.

The conditions of processing, packing, or storage also may be factors in a_w control in foods. Temperature is very important in this respect. The degree of solubility of many humectants is affected by temperature and, although a liquid system containing sucrose as a humectant may be saturated at process temperatures near ambient, if subsequently refrigerated, some sucrose may crystallize and thus increase the measured a_w. This change may be sufficient to allow molds to grow or otherwise adversely affect the properties of a food. In situations where this is a problem, the addition of certain gums or thickening agents may be useful in preventing precipitation by forestalling crystal initiation.

a_w Adjustment Methods

A number of processes and procedures have been utilized throughout history to provide adjustment or control of a_w levels in the laboratory.

Gas Streams

Streams of gas adjusted to various humidity levels may be directed through a food to adjust it to a predetermined a_w level. This method of a_w control has utility primarily with dry, particulate foods through which the flowing gas percolates.

Desired relative humidity levels are attained in gases by mixing "dry" and "wet" air streams (Smith, 1965), passage of air through H_2SO_4 solutions of known a_w (Sair and Fetzer, 1944), or passage through saturated salt solutions of known a_w (Ayerst, 1965b). Bubbling large volumes of air through less-than-saturated salt solutions has the advantage that predetermined a_w gradations can be "targeted" to desired intervals. The continual removal of water by the air stream, however, tends to increase the salt concentration in the solution and, hence, to lower a_w. Additional methods employing gases to adjust a_w may be found in the review edited by Wexler and Brombacher (1951).

Salts

Solutions of salts, saturated or unsaturated, organic or inorganic, are frequently used in laboratories to obtain desired a_w levels. These solutions may be used to control a_w in gas streams (as above), as standard solutions for the calibration of a_w-measuring instruments, or as equilibrating solutions for small food samples. In the latter case, the food is equilibrated to a predetermined a_w

level by enclosure in a chamber, usually a desiccator, with an appropriate saturated salt or a solution containing an amount of salt to produce the desired a_w. Often, equilibration rates within such chambers are very slow, and so precautionary measures may be required to prevent microbial growth in food samples at a_w levels > 0.70. Increased equilibration rates may be obtained by evacuation or by atmosphere movement through the use of fans or blowers mounted within the chamber. Equilibration time also may be shortened by exposing a maximal surface of the food to the chamber atmosphere through grinding and/or spreading to a thin layer. A graph relating a_w to the concentration of a salt, NaCl, in distilled water is shown in Fig. 2.13.

Saturated solutions are preferred to unsaturated solutions for most equilibration applications because the need to measure precise quantities of the humectant has been omitted, and excess salt crystals in the solution act as a "buffer" to prevent a_w alteration as the solution absorbs water from the enclosed atmosphere (desorbing mode). Difficulties arise, however, in determining the absolute a_w levels of such solutions. Numerous literature reports provide listings (Young, 1967; Winston and Bates, 1960; Rockland, 1957; Stokes and Robinson, 1949); however, as noted earlier in this chapter, agreement among these tables may not be precise. A comparison of some examples of these data is shown in Table 2.1. The publications of Sloan and Labuza (1975b), Carr and Harris (1949), and Labuza *et al.* (1976) provide additional information on the a_w of saturated salts.

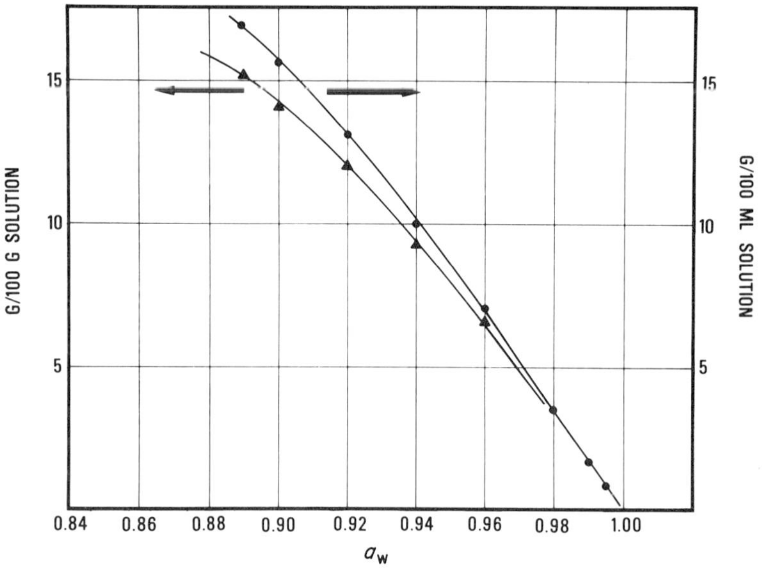

Fig. 2.13. Water activity of NaCl solutions in distilled water at 25°C.

In addition, the publication of Winston and Bates (1960) presents the E. R. H. values for nearly 100 compounds at 11 temperatures ranging from 2° to 50°C. These values were obtained by interpolative methods.

Mixtures of saturated salts also may be employed to produce a_w levels lower than those obtained with either salt alone. For example, Winston and Bates (1960) combined saturated solutions of NaCl (a_w = 0.755) and KCl (a_w = 0.850) to produce a mixture (presumably 1:1) of 0.715 a_w. Other pairs of solutes are listed in this publication; however, the authors caution that only cations or anions common to both salts give consistent results. In addition to chemical exchange reactions between salt pairs, the formation of complex radicals may alter vapor pressures of such mixtures in an unpredictable manner.

Sulfuric Acid Solutions

Control of a_w also may be achieved by allowing a sample to equilibrate with known solutions of H_2SO_4. Table 2.3 lists a_w values that can be attained in materials at equilibrium with such solutions. The usual laboratory safety precautions apply to the preparation and handling of strong acid solutions.

Multicomponent Equilibration within a Food

The a_w of a food can often be controlled by the addition of components or ingredients of different a_w. The addition of an inorganic salt such as NaCl is an example already discussed; however, food components themselves may serve in a similar capacity. In these circumstances, a food must be equilibrated internally during processing and subsequent storage. If this equilibration occurs slowly, a potential exists for microbial growth or other deleterious alterations that might occur before the a_w reaches a range which is inhibitory to such changes.

Simple removal of water is, of course, the most common method of controlling a_w in foods. These methods are covered in numerous volumes devoted specifically to this subject (Van Arsdell *et al.*, 1973; von Loesecke, 1955). Water addition can be achieved by mixing; however, in the case of dry products this may be complicated by caking with resulting poor moisture distribution. Steam injection of high humidity gas streams finds some application in these instances. Davies *et al.* (1969) have described a process in which water is added to dry foods by sublimation from powdered ice at subfreezing temperatures. Using this procedure, these authors were able to adjust wheat flours within a range of 14–60% water.

Organic Compounds

Sucrose, glycerol, and propylene glycol are three organic compounds often used to lower a_w levels in foods. Other sugars such as levulose, maltose, etc. have also found some limited use, and invert sugar (1:1 glucose/levulose mixture) is

TABLE 2.4.

Glycerol Solutions for Control of Relative Humidity[a]

a_w (25°C)	Refractive index	Percent glycerol (by wt.)	Grams glycerol per liter solution
0.98	1.3463	11.25	—
0.96	1.3560	18.80	—
0.95	1.3602	22.0	231.50
0.90	1.3773	34.90	378.92
0.85	1.3905	44.72	497.33
0.80	1.4015	52.30	592.34
0.75	1.4109	58.61	675.14
0.70	1.4191	64.15	747.43
0.65	1.4264	69.05	813.89
0.60	1.4329	73.40	873.86
0.55	1.4387	77.30	928.46
0.50	1.4440	80.65	976.03
0.40	1.4529	86.30	—

[a] Grover and Nicol, 1940.

an important agent in some confectionery applications. With the possible exception of NaCl (discussed above), sucrose has probably received broader and longer application than any other humectant. The use of sucrose is limited primarily to jellies, syrups, fondants, and other confectionlike materials in which its sweet flavor is acceptable. A relatively new application is in intermediate moisture dog food meat or meatlike patties in the United States. Apparently, dogs do not object to sugar in their meat (Kaplow, 1970) to the extent that humans probably would.

Rather high concentrations of glycerol are required to produce even a relatively modest reduction in a_w. Table 2.4 lists the a_w values obtained (25°C) with various glycerol–water solutions. Flavor considerations often rule out application of this compound in foods since it adds a "spicy" or "hot" note to foods. In laboratory studies, the use of glycerol to adjust a_w levels in the microbial growth range can create errors since many bacteria can metabolize glycerol (and/or sucrose) with the possibility that an upward adjustment of a_w occurs. Also, many microorganisms tend to be more tolerant of a_w depression if glycerol is employed to obtain the reduction.

Propylene glycol also has received limited application as a food humectant. This compound has the advantage that it possesses intrinsic antimicrobial properties in addition to its activity as a humectant (Haas *et al.*, 1975); however, it may be toxic at use concentrations. A patent has been granted (Frankenfeld, *et al.*, 1973) for the use of 1,3-diols, such as 1,3-butanediol, to adjust a_w levels in intermediate moisture foods. Some antimicrobial properties are also accorded this compound. The properties of these and other humectants in food applications have been reviewed by Sloan and Labuza (1975a).

Further discussion of techniques used commercially or in laboratories to reduce a_w can be found in Chapters 9 and 10.

REFERENCES

Anagnostopoulos, G. D. (1973). Water activity in biological systems: A dew point method for its determination. *J. Gen. Microbiol.* **77,** 233–235.

A.O.A.C. (1975). "Methods of Analysis," 11th ed. Assoc. Off. Anal. Chem., Washington, D.C.

Ayerst, G. (1965a). Determination of the water activity of some hygroscopic food materials by a dew point method. *J. Sci. Food Agric.* **16,** 71–78.

Ayerst, G. (1965b). Water activity—its measurement and significance in biology. *Int. Biodeterior. Bull.* **1,** 13–26.

Blaine, R. L. (1975). Thermal analysis for rapid, precise testing of food ingredients, packaging materials, *Food Prod. Dev.* **9,** 30–33.

Bloch, F., Hansen, W. W., and Packard, M. F. (1946). The nuclear induction experiment. *Phys. Rev.* **70,** 474–485.

Bosin, W. A., and Easthouse, H. D. (1970). Rapid method for obtaining humidity equilibrium data. *Food Technol. (Chicago)* **24,** 1155–1178.

Brastad, W. A., and Borchardt, L. F. (1953). Electric hygrometer of small dimensions. *Rev. Sci. Instrum.* **24,** 1143–1144.

Brekke, J., and Conrad, R. (1965). Gas–liquid chromatography and vacuum oven determination of moisture in fruits and fruit products. *J. Agric. Food Chem.* **13,** 591–593.

Carr, D. S., and Harris, B. L. (1949). Solutions for maintaining constant relative humidity. *Ind. Eng. Chem.* **41,** 2014–2015.

Cutting, C. L., Jason, A. C., and Wood, J. C. (1955). A capacitance–resistance hygrometer. *J. Sci. Instrum.* **32,** 425–431.

d'Alton, G. (1969). A graphical interpolation method. *Confect. Manuf. Market.* **3,** 184–186.

Davies, R. J., Daniels, N. W. R., and Greenshields, R. N. (1969). An improved method of adjusting flour moisture in studies on lipid binding. *J. Food Technol.* **4,** 117–123.

Duckworth, R. B. (1971). Differential thermal analysis of food systems. I. The determination of unfreezable water. *J. Food Technol.* **6,** 317–327.

Fett, H. M. (1973). Water activity determination in foods in the range 0.80 to 0.99. *J. Food Sci.* **38,** 1097–1098.

Fischer, K. (1935). A new method for the analytical determination of the water content of liquids and solids. *Angew. Chem.* **48,** 394–396.

Fosnot, R. H., and Haman, R. W. (1945). A preliminary investigation of the application of Karl Fischer reagent to determination of moisture in cereals and cereal products. *Cereal Chem.* **22,** 41–49.

Frankenfeld, J. W., Karel, M., and Labuza, T. P. (1973). Intermediate moisture food compositions containing aliphatic 1,3-diols. U.S. Patent 3,732,112.

Gough, M. C. (1974). The measurement of relative humidity with particular reference to remote long-term measurement in grain silos. *Trop. Stored Prod. Inf.* **27,** 19–30.

Grover, D. W., and Nicol, J. M. (1940). The vapor pressure of glycerin solutions at 20°C. *J. Soc. Chem. Ind., London* **59,** 175–177.

Gur-Arieh, C., Nelson, A. I., Steinberg, M. P., and Wei, L. S. (1965). A method for rapid determination of moisture—adsorption isotherms of solid particles. *J. Food Sci.* **30,** 105–110.

Haas, G., Bennett, J. D., Herman, E. B., and Collette, D. (1975). Microbial stability of intermediate moisture foods. *Food Prod. Dev.* **9,** 86–90 and 94.

Jason, A. C. (1965). Some properties and limitations of the aluminum oxide hygrometer. *In* "Humidity and Moisture" (A. Wexler and R. E. Ruskin, eds.), Vol. I, pp. 372–390. Van Nostrand-Reinhold, Princeton, New Jersey.

Kaplow, M. (1970). Commercial development of intermediate moisture foods. *Food Technol.* **24,** 889–893.

Karan-Djurdjic, S., and Leistner, L. (1970). Measurement of the water activity of meat and meat products. *Fleischwirtschaft* **50,** 1104–1106.

Karel, M., and Nickerson, J. T. R. (1964). Effects of relative humidity, air and vacuum on browning of dehydrated orange juice. *Food Technol. (Chicago)* **18,** 1214–1218.

Kobayashi, J., and Toyama, Y. (1965). The aging effect of an electrolytic hygrometer. *In* "Humidity and Moisture" (A. Wexler and R. E. Ruskin, eds.), Vol. I, pp. 248–264. Van Nostrand-Reinhold, Princeton, New Jersey.

Labuza, T. P., Lee, R. Y., Flink, J., and McCall, W. (1976). Water activity determination: A collaborative study of different methods. *J. Food Sci.* **41,** 910–917.

Landrock, A. H., and Proctor, B. E. (1951). Measuring humidity equilibria. *Mod. Packag.* **24,** 123–130 and 186.

Little J. W., Hasegawa, S., and Greenspan, L. (1974). "Performance Characteristics of a 'Bulk Effect' Humidity Sensor," Bull. 74-477. Natl. Bur. Stand., Washington, D.C.

Loncin, M. (1955). The solubility of water in fats and oils and the vapor pressure of the dissolved water. *Fette, Seifen, Anstrichm.* **57,** 413–415.

Makower, B., and Meyers, B. (1943). A new method for the determination of moisture in dehydrated vegetables. *Proc. Inst. Food Technol.,* 156–164.

Mallett, D., Kohnen, J. B., and Surles, T. (1974). Determination of water activity of intermediate moisture pet foods by solvent extraction. *J. Food Sci.* **39,** 847–848.

Miller, B. S., and Kaslow, H. D. (1963). Determination of moisture by nuclear magnetic resonance and oven methods in wheat, flour, doughs and dried fruits. *Food Technol. (Chicago)* **17,** 650–653.

Miyata, A., and Watari, H. (1965). A hygrometer which utilizes an anodic oxide film of aluminum. *In* "Humidity and Moisture" (A. Wexler and R. E. Ruskin, eds.), Vol. I, pp. 391–404. Van Nostrand-Reinhold, Princeton, New Jersey.

Mossel, D. A. A., and van Kuijk, H. J. L. (1955). A new and simple technique for the direct determination of the equilibrium relative humidity of foods. *Food Res.* **20,** 415–423.

Oxley, T. A., and Pixton, S. W. (1961). Determination of moisture content in cereals. II. Errors in determination by oven drying of known changes in moisture content. *J. Sci. Food Agric.* **11,** 315–319.

Palmer, K. J., and Elsken, R. H. (1956). Determination of water by nuclear magnetic resonance in hygroscopic materials containing soluble solids. *J. Agric. Food Chem.* **4,** 165–167.

Parducci, L. G., and Duckworth, R. B. (1972). Differential thermal analysis of frozen food systems. II. Micro-scale studies on egg white, cod and celery. *J. Food Technol.* **7,** 423–430.

Pomeranz, Y., and Lindner, C. (1953). A simple method for the evaluation of moisture in milled products. *Bull. Res. Counc. Isr.* **3,** 251.

Pomeranz, Y., and Meloan, C. E. (1971). "Food Analysis: Theory and Practice." Avi Publ. Co., Westport, Conn.

Purcell, E. M., Torey, H. C., and Pound, R. V. (1946). Resonance adsorption by nuclear magnetic moments in a solid. *Phys. Rev.* **69,** 37–38.

Robinson, R. A., and Sinclair, D. A. (1934). The activity coefficients of the alkali chlorides and of lithium chloride in aqueous solution from vapor pressure measurements. *J. Am. Chem. Soc.* **56,** 1830–1835.

Robinson, R. A., and Stokes, R. H. (1959). "Electrolyte Solutions." Butterworth, London.

Rockland, L. B. (1957). A new treatment of hygroscopic equilibria: Application to walnuts (Juglans regia) and other foods. *Food Res.* **22,** 604–628.

Rockland, L. B. (1960). Saturated salt solutions for static control of relative humidity between 5° and 40°C. *Anal. Chem.* **32,** 1375–1376.

Rodel, W., and Leistner, L. (1972). Messung der Wasseraktivität (a_w-Wert) von Fleisch und Fleischwaren mit einem Taupunkt-Hygrometer. *Fleischwirtschaft* **52,** 1461–1462.

Sair, L., and Fetzer, W. R. (1944). Water sorption by starches. *Ind. Eng. Chem.* **36,** 205–208.

Scatchard, G., Hamer, W. J., and Wood, S. E. (1938). Isotonic solutions. I. Chemical potential of water in aqueous solutions of sodium chloride, potassium chloride, sulfuric acid, sucrose, urea and glycerol at 25°. *J. Am. Chem. Soc.* **60,** 3061–3070.

Schwecke, W. M., and Nelson, J. H. (1964). Determination of moisture in foods by gas chromatography. *Anal. Chem.* **36,** 689–690.

Shaw, T. M., and Elsken, R. H. (1956). Determination of water by nuclear magnetic absorption in potato and apple tissue. *J. Agric. Food Chem.* **4,** 162–164.

Shaw, T. M., Elsken, R. H., and Kunsman, C. H. (1953). Moisture determinations of foods by hydrogen nuclei magnetic resonance. *J. Assoc. Off. Agric. Chem.* **36,** 1070–1076.

Sloan, A. E., and Labuza, T. P. (1975a). Humectant water sorption isotherms. *Food Prod. Dev.* **9,** 68.

Sloan, A. E., and Labuza, T. P. (1975b). Investigating alternative humectants for use in foods. *Food Prod. Dev.* **9,** 75, 78, 80, 82, 84, and 88.

Smith, P. R. (1965). A new apparatus for the study of moisture sorption by starches and other foodstuffs in humidified atmospheres. *In* "Humidity and Moisture" (A. Wexler and W. A. Wildhack, eds.), Vol. III, pp. 487–494. Van Nostrand-Reinhold, Princeton, New Jersey.

Smith, P. R. (1971). The determination of equilibrium relative humidity or water activity in foods—a literature review. *B.F.M.I.R.A. Sci. Tech. Surv.* No. 70.

Solomon, M. E. (1957). Estimation of humidity with cobalt thiocyanate papers and permanent colour standards. *Bull. Entomol. Res.* **48,** 498–506.

Sood, V. C., and Heldman, D. R. (1974). Analysis of vapor pressure manometer for measurement of water activity in non-fat dry milk. *J. Food Sci.* **39,** 1011–1013.

Stokes, R. H. (1947). The measurement of vapor pressures of aqueous solutions by bithermal equilibration through the vapor phase. *J. Am. Chem. Soc.* **69,** 1291–1296.

Stokes, R. H., and Robinson, R. A. (1949). Standard solutions for humidity control at 25°C. *Ind. Eng. Chem.* **41,** 2013.

Sudhakar, S., Steinbert, M. P., and Nelson, A. I. (1970). Bound water defined and determined at constant temperature by wide-line NMR. *J. Food Sci.* **35,** 612–615.

Swift, J. R. (1971a). Moisture measurements in the food industry, an infrared absorption technique. *Food Trade Rev.* **41,** 25–26.

Swift, J. R. (1971b). Moisture measurements in the food industry. An infrared technique. *Food Technol. Aust.* **23,** 352–353.

Szulmayer, W. (1969). Humidity and moisture measurement. *CSIRO Food Preserv. Q.* **29,** 27–35.

Taylor, A. A. (1961). Determination of moisture equilibria in dehydrated foods, *Food Technol. (Chicago)* **15,** 536–540.

Toledo, R. (1973). Determination of water activity in foods. *Proc. Meat Ind. Res. Conf.* pp. 85–106.

Toledo, R., Steinberg, M. P., and Nelson, A. I. (1968). Quantitative determination of bound water by NMR. *J. Food Sci.* **33,** 315–317.

Troller, J. A. (1977). Statistical analysis of a_w measurements obtained with the Sina Scope. *J. Food Sci.* **41,** 86–90.

Van Arsdell, W. B., Copley, M. J., and Morgan, A. I., Jr. (1973). "Food Dehydration," Vol. I. Avi Publ. Co., Westport, Connecticut.

von Loesecke, H. W. (1955). "Drying and Dehydration of Foods," 2nd ed. Van Nostrand-Reinhold, Princeton, New Jersey.

Vornheder, P. F., and Brabbs, W. J. (1970). Moisture determination by near-infrared spectrometry. *Anal. Chem.* **42,** 1454–56.

Vornhof, D. W., and Thomas, J. H. (1970). Determination of moisture in starch hydrolysates by near-infra-red and infra-red spectrophotometry. *Anal. Chem.* **42,** 1230–1233.

Vos, P. T., and Labuza, T. P. (1974). Technique for measurement of water activity in the high a_w range. *J. Agric. Food Chem.* **22,** 342–343.

Ward, C. B., and Tischer, R. G. (1953). Use of cobaltous chloride to detect moisture patterns in partially dehydrated kernels of corn. *Cereal Chem.* **30,** 420–426.

Washburn, E. W., ed. (1926). "International Critical Tables of Numerical Data, Physics, Chemistry and Technology," 1st ed. McGraw-Hill, New York.

Weast, R. C., ed. (1972–1973). "Handbook of Chemistry and Physics," 53rd ed., p. 40. Chem. Rubber Publ. Co., Cleveland, Ohio.

Weise, E. L., Burke, R. W., and Taylor, J. K. (1965). I. Gas chromatographic determination of the moisture content of grain. *In* "Humidity and Moisture" (A. Wexler and P. N. Winn, eds.), Vol. IV, pp. 3–6. Van Nostrand-Reinhold, Princeton, New Jersey.

Wendlandt, W. W. (1974). "Thermal Methods of Analysis," 2nd ed. Wiley, New York.

Wernimont, G., and Hopkinson, F. J. (1943). The dead-stop end point. As applied to the Karl Fischer method for determining moisture. *Anal. Chem.* **15,** 272–273.

Wexler, A., ed. (1965). "Humidity and Moisture," Vols. I–IV. Van Nostrand-Reinhold, Princeton, New Jersey.

Wexler, A., and Brombacher, W. G. (1951). Methods of measuring humidity and testing hygrometers. *Natl. Bur. Stand. (U.S.), Circ.* **512.**

Wexler, A., and Hasegawa, S. (1954). Relative humidity–temperature relationships of some saturated salt solutions in the temperature range 0° to 50°C. *J. Res. Natl. Bur. Stand.* **53,** 19–26.

Winston, P. W., and Bates, D. H. (1960). Saturated solutions for the control of humidity in biological research. *Ecology* **41,** 232–237.

Wodzinski, R. J., and Frazier, W. C. (1960). Moisture requirements of bacteria. I. Influence of temperature and pH on requirements of *Pseudomonas fluorescens*. *J. Bacteriol.* **79,** 572–578.

Young, J. F. (1967). Humidity control in the laboratory using salt solutions—a review. *J. Appl. Chem.* **17,** 241–245.

3

Enzyme Reactions and Nonenzymatic Browning

EFFECT OF WATER ACTIVITY ON ENZYMATIC REACTIONS

The diversity and high degree of specific reactivity that characterize enzymes, in general, also characterize their action in foods. Enzymes may be added to foods intentionally to catalyze desired alterations, e.g., the addition of chill-proofing agents (proteolytic enzymes) to beer to prevent the precipitation of proteins when it is refrigerated. A similar type of enzyme is used to achieve a very different objective when certain meats are treated with protein-digesting enzymes to increase tenderness and, thereby, consumer acceptance. Beneficial reactions can also be carried out by enzymes developed in foods through the growth of specific types of microorganisms. For example, there are the proteases and lipases synthesized by the mold, *Penicillium roqueforti,* that produce desirable flavors during the ripening of Roquefort or blue cheese.

Equally important are those detrimental changes that are mediated by enzymes occurring naturally in foods, such as the nonmicrobial decomposition of fruits and vegetables during handling and storage. Many food processes include steps designed specifically to thwart the activity of these naturally occurring enzymes in foods. An important example is the blanching of fruits and vegetables by hot water or steam to inactivate enzymes and thus delay or prevent deleterious changes. Other types of detrimental enzymatic reactions are the browning reactions in fruits and vegetables catalyzed by phenol oxidases and rancidity in flour resulting from the activities of lipases and lipoxidases occurring naturally in wheat germ.

Enzymes may be defined as proteinaceous compounds that catalyze organic reactions in a highly specific and reactive manner. The increase in reaction rate is generally 4–5 orders of magnitude in an enzymatically catalyzed reaction, but in some cases it may be even greater. This catalysis is characterized by two general processes: (1) the binding of enzyme to the substrate, usually as a noncovalent complex or as an acyl-enzyme intermediate; and (2) the breakdown of this complex into a reaction product(s) and the enzyme. Although little has been published in the food-related literature concerning the part played by water in these processes, it is probably involved in both stages of the enzymatic reaction with its major function being to increase the mobility of substrate and the reaction product(s). In addition, in enzyme-mediated hydrolytic reactions, such as the example described in equation (1),

(1)

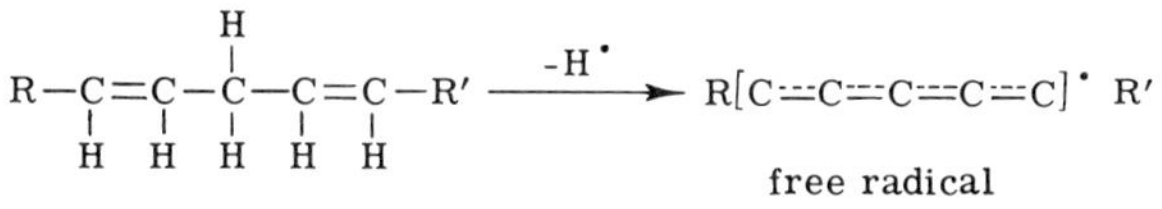

water participates actively in the reactions. The actual amount of water utilized during hydrolysis depends on many variables, including the amount and nature of the substrate and enzymes. Reactions of this type are probably not limited in low-moisture foods as a result of insufficient "participating" moisture, since many enzymatic reactions cease at water concentrations well in excess of those theoretically required for hydrolysis (Jencks, 1969).

Probably little water is available for the movement of substrate and products at a_w levels below the BET monolayer (see Chapter 1). At these a_w levels, freely mobile water is not available to carry out reactions, and so enzymatic reactions tend to be suppressed in the lower regions of the sorption isotherm (Fig. 3.1). However, as the a_w of a food is increased, capillary condensation commences, and enzymatic reaction rates increase. Acker (1962) has described this as a process in which increasingly large capillary pores are filled with water, resulting in greater substrate dissolution and increasing reaction rates. Obviously, the capillarity of a given food must play a vital role in defining the minimal a_w at which an enzymatic process can occur. A demonstration of the effect that capillarity exerts on the activity of enzymes was obtained by Kiermeier and Corduro (1954), who tested starch/amylase systems and found no amylase activity in preparations containing 14% water. However, when the reaction was carried out on filter papers, hydrolysis occurred in the presence of only 10% water. The observed differences in these experiments are related to the availability of free water, which tends to persist at low relative humidities in the pores of the filter paper. In this context, the conformation of the hysteresis loop, which predicts the presence of free water in a food, can be extremely useful in estimating the rate and relative extent of enzymatic reactions (Rockland, 1957; Acker, 1963).

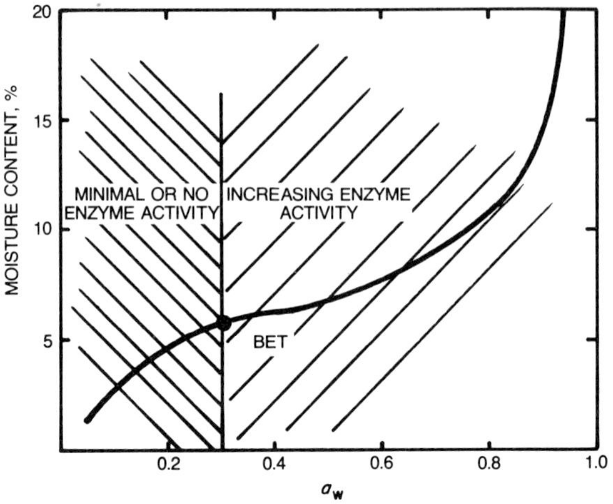

Fig. 3.1. Relationship between the hysteresis curve and enzyme activity in a typical reaction system.

Enzyme Stability

The stability of enzymes is often critical in determining whether a food processor can utilize added enzymes to obtain a desired effect in a food or to alter a processing step to inactivate undesirable enzymes. Most commercial, food-grade enzyme preparations are sold on the basis of units of activity, and so it is clearly desirable to maintain their potency during manufacture and storage. This is a particular problem with liquid enzyme products. To stabilize these preparations, compounds such as sodium chloride, glycerol, or propylene glycol may be added. These materials, by reducing the a_w level, effectively "remove" water and, in so doing, increase the stability of the enzyme(s) present. Food enzymes normally retain their activity for extended periods if maintained at near neutral pH levels, moderate or low temperatures, and reduced moisture. The latter condition not only retards denaturation of the enzyme (and thus its activity) but also prevents the growth of microorganisms that might metabolize it. Under most circumstances, an a_w level of 0.30 or lower will prevent such deterioration, yet will preserve much of the original enzymatic activity. Acker and Kaiser (1959), in studies on the hydrolysis of lecithin, showed that barley lecithinase retained its activity at 30°C for at least 48 days when the system was maintained at an a_w level of 0.35 (Fig. 3.2). Subsequent addition of water to raise the a_w level to 0.70 produced hydrolysis comparable to that of a control system held at the higher a_w. In considering the effects of a_w on enzyme storage, one should recognize that different enzymes react in very dissimilar fashions, so that conditions which stabilize one enzyme may lead to the deterioration of another. Additionally, a

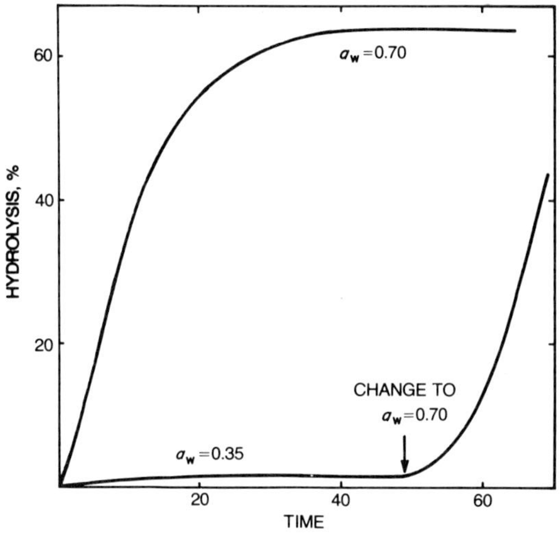

Fig. 3.2. Lecithin hydrolysis at 0.70 and 0.35 a_w (Acker and Kaiser, 1959).

number of factors other than a_w (Table 3.1) may act or interact to influence enzymatic reactions in foods.

From the processing standpoint, the heat stability of enzymes at various moisture contents is probably more important than the stability of enzyme preparations for extended storage periods at ambient temperatures. Taken alone, heat may destabilize an enzyme by several possible mechanisms, and occasionally heat inactivation may be reversible, as is the case with peroxidase in milk (Reed, 1966), and spinach catalase (Sapers and Nickerson, 1962). Polyhydric alcohols, such as sorbitol (70%), have been found (Yasumatsu *et al.*, 1965) to stabilize α-amylase and an alkaline protease against denaturation by heating at 80°C. Solutions of 70% glycerol, 70% sucrose, and 70% glucose were also reported

TABLE 3.1.

Factors Affecting the Stability of Enzymes in Foods

1. Temperature
2. pH
3. Ionic strength
4. Moisture level
5. Nature of food
6. Time of storage
7. Presence of activators and inhibitors

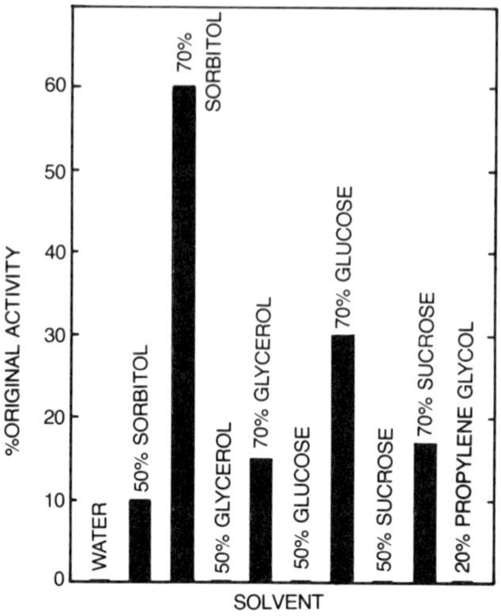

Fig. 3.3. Effects of various solutes on the thermal (80°C, 5 minutes) stability of crystalline protease (Yasumatsu *et al.*, 1965).

(Fig. 3.3) to be effective in suppressing the thermal inactivation of the proteinase. These authors also showed (Fig. 3.4) that 70% sorbitol effectively protected α-amylase, although no protection was apparent until the concentration reached at least 50%. Glycerol was somewhat less protective.

The mechanism by which enzymes are stabilized against heat in low moisture environments is not fully understood. In the absence of water-limiting solutes,

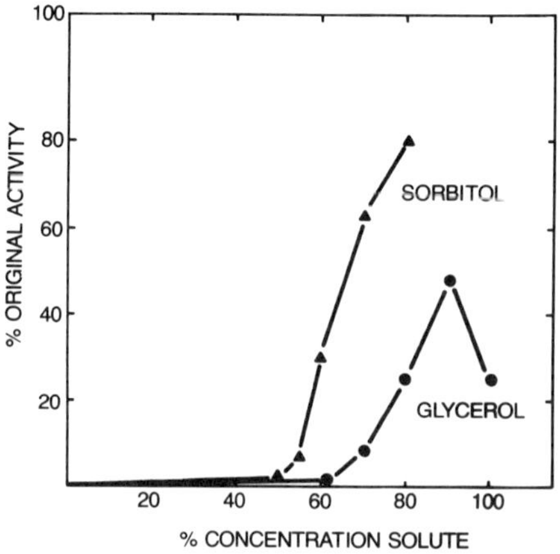

Fig. 3.4. Effects of 70% sorbitol and 70% glycerol on the activity of α-amylase on soluble starch at various solute concentrations. (Heated at 80°C for 5 minutes.) (Yasumatsu *et al.*, 1965.)

thermal inactivation usually follows first-order reaction kinetics, and this situation probably also prevails at reduced a_w levels. During heating, some rupture of hydrophobic bonds occurs, which alters the configuration of the molecule, a process often referred to as unfolding. It is this unfolding which is largely responsible for the loss in enzyme activity observed in heated solutions, and therefore, the stability conferred by high ionic strength systems may be the result of a stabilization of hydrophobic bonds. With further heating, some denaturation of the enzyme's active sites may occur, but it is doubtful whether the peptide "backbone" of the enzyme is cleaved unless the heating conditions are severe or occur at pH levels in excess of 8.0.

Model Systems

In many foods, a variety of chemical compounds may be present that mask and influence enzymatic reactions and make the reaction products difficult to detect. Thus, model systems studies utilizing relatively simple substrates have been of value in predicting the extent of these reactions in actual food systems. Occasionally, the extrapolation required for these comparisons is great, and so accuracy may suffer; however, it is advantageous to have knowledge of a given reaction occurring under ideal conditions with known ingredients.

An example of a model system study is that carried out by Blaine (1962) using water/glycerol solutions to control the moisture level of reaction systems contain-

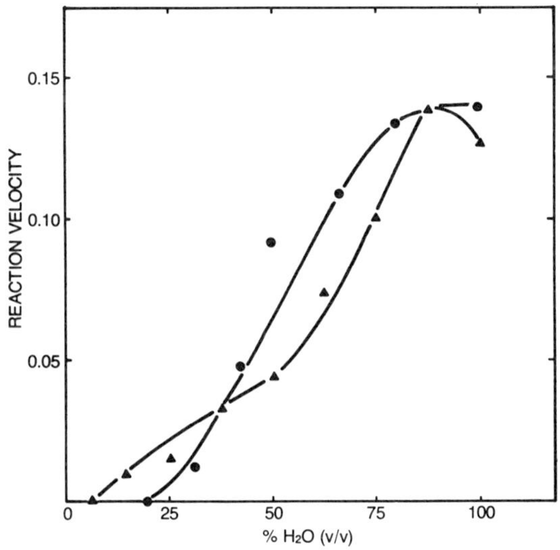

Fig. 3.5. The effect of water/glycerol mixtures on reaction velocities of peroxidase (△) and lipoxidase (●) (Blaine, 1962).

ing two oxidases—peroxidase and a crude lipoxidase from soybean extracts. A slight stimulation of peroxidase activity was noted in the presence of small amounts of glycerol (Fig. 3.5). With increases in glycerol content in excess of 12–13% (i.e., 87–88% water; about 0.975 a_w), a marked decline in activity was noted until at 90–92% glycerol (8–10% H_2O; $a_w = 0.230$), enzymatic activity could no longer be detected. When either dioxane or dimethylsulfoxide was substituted for glycerol at equivalent percent concentrations (but not a_w levels), a demonstrable decrease in enzyme activity occurred. In the case of lipoxidase, the reaction ceased at 20% water and did not exhibit an optimal "peak" at very low glycerol concentrations, as did peroxidase. However, the lipoxidase did reach and maintain a maximum reaction velocity at a lower water content than did the peroxidase.

Other reports of investigations of model enzyme systems, not discussed in detail here, have all shown that the amount of moisture present, its degree of binding to the substrate, and the nature of the substrate are important factors in determining reaction velocities. However, many of these studies suffer from the lack of clearly defined reaction criteria, e.g., a failure to define the homogeneity and purity of the enzyme preparations and to treat the data obtained in terms of enzyme kinetics and simple rate theory. Until data meeting these criteria are available, generalizations on the effects of a_w limitation on enzymes or types of enzymes are hazardous. For this reason, examples of various enzymatic reactions have been chosen to acquaint the reader with the moisture conditions generally required for the individual reactions to occur in the particular food under investigation. Where possible, the a_w equivalents of the moisture contents mentioned in the original publications have been included. The selection of enzyme type, although arbitrary, is based largely on the importance of the reaction in producing deleterious or beneficial changes in foods.

Lipases

Lipases catalyze hydrolysis of triglycerides with the production of glycerol, fatty acids, and di- and monoglycerides. Many foods, including cereal grains and milk, contain endogenous levels of this enzyme type that react with food constituents to produce off-flavors. Exogenous lipases, such as those produced by molds growing in a food, may also create problems by the formation of "rancid" flavors; although, in the case of some types of cheese, the flavors produced are highly desirable.

Increases in the acidity of dehydrated meats containing 3.2% water have been attributed to the action of endogenous lipases. Cereal products containing from 6 to 20% moisture are similarly subject to the action of these enzymes. Usually, the initial reaction velocities increase dramatically with increases in moisture content, e.g., Rothe (1958) noted that an increase from 8.8 to 15.1% moisture in wheat flour produced a fivefold increase in lipase activity. Some chemical bleaching or whitening processes may not completely inactivate lipases in wheat flour, although baking will usually inactivate, or at least drastically reduce the activity of these enzymes. Lipolytic activity, as demonstrated by "fat acidity" in flours (Cuendet *et al.*, 1954), is normally related to the degree of flour refinement and moisture level, and may manifest itself in finished bread as off-, or rancid odors and flavors and in reduced loaf volume. In flours of varying degrees of refinement, no lipase activity was detected at 3% moisture levels over a 52-week storage period, whereas at 14% H_2O, acidity developed rapidly and to relatively high levels. In most types of flours, the quantity of substrate available for reaction becomes limiting within 20–25 weeks at the higher moisture level. The activity of lipase can be particularly troublesome in foods such as cake mixes, in which shortening and lipase-containing flours may come into contact in the presence of sufficient moisture (7.4%) to permit lipolysis (Matz *et al.*, 1955). High-grade flours currently used in cake mixes are thoroughly refined or milled, a process which results in the removal of the lipase-containing germ. This factor, plus the normally low a_w of these products (0.17–0.20), prevents acidic off-odors.

The curing and flavor development of certain cheeses through the action of mold-produced lipases were noted earlier in this chapter. Fungi may also secrete lipases in cereal products, with the production of sour flavors. Normally, molds will not grow in grains with moisture contents of less than 12% (Snow *et al.*, 1944) to 13.5% (Poisson and Guilbot, 1963). However, if these products have been held at higher ambient humidities after harvest, followed by drying to 7–10% moisture, lipases secreted during the period of active mold growth may react subsequently with the triglycerides present in grains to produce fatty acids. Because molds also produce organic acids, such as citric acid, as metabolic products, the nature and thus the origin of acidic off-flavors is difficult to determine unless careful extraction and analytical techniques (such as vapor-phase chromatography) are employed.

In model systems studies, the effect of moisture level on lipase activity is not as clear. For example, Purr (1966) demonstrated that both mold and pancreatic lipases hydrolyzed "hard" and "soft" coconut fats at very low humidities, even below the BET monolayer levels of these substrates. Thus, it seems plausible that lipolysis is not dependent on free water. There have also been reports (Caillot and Drapron, 1970; Acker and Beutler, 1963) of lipase activity at a_w levels as low as 0.025. The substrate in these systems was triolein, which, at the temperatures of incubation used in these studies, is a liquid and thus able to diffuse toward the enzyme in the absence of appreciable free water (Drapron, 1972). Based on these results, one could assume that both substrate and reaction product mobility are key factors and that water usually serves merely as a convenient means of providing the mobility in the case of water-soluble or water-miscible substrates or products.

Further discussions on the mechanisms by which water activity influences the activity of lipases can be found in the interesting publication of Drapron (1972). The principal thrust of that work lies in the hypothesis that at low a_w levels enzyme hydration is insufficient to permit the enzyme molecule to assume the optimal configuration for reactivity. Insofar as the present authors are aware, this is the only reference to possible enzyme configurational effects, and so the hypothesis put forward by Drapron deserves further investigation.

Guardia and Haas (1967) determined the effect of water-binding solutes on the activity of pancreatic, mold, and plant (wheat germ) lipases. The three solutes tested, sucrose, glycerol, and propylene glycol,inhibited pancreatic lipase activity at pH 9.0, but stimulated it at pH 7.0. A fungal lipase, however, was strongly inhibited by as little as 10% (v/v) propylene glycol at pH 7.0 and virtually inactivated in the presence of 40% of this compound. In investigations on the a_w effects of solutes during heating, sucrose and glycerol significantly protected pancreatic lipase, whereas propylene glycol increased its thermolability. The observed enhancement of pancreatic lipase activity described above could not be attributed to emulsion particle size or interactions with emulsifiers, nor could it be related to water content of the enzyme systems. Unfortunately, the equilibrium relative humidities noted in these studies do not agree with values published elsewhere for similar concentrations of propylene glycol and glycerol. Despite this difficulty, the apparent stimulation of lipase activity by the three solutes, while unexplained, could be of significance in foods in which a_w levels are controlled by these materials. Similarly, the protection of this enzyme by sucrose and glycerol solutions during heating could play an important role in food processes that include blanching or other enzyme-inactivating steps.

Proteolytic Enzymes

Proteases occur as constituents of many foods and find use in products as diverse as beer (chillproofing), meats (tenderizers), and bread doughs (reduction of dough strength to reduce mixing times).

Many types of proteases, and especially fungal proteases, are strongly inhibited by 2–3% NaCl. In sponge bread doughs, however, it is doubtful if this inhibition can be attributed to the small reduction in water activity that these concentrations of NaCl would produce, even if one considers that only 25% of the water in these doughs is "free" and, hence, available as a solvent for the NaCl. On the other hand, liquid meat-tenderizing preparations, usually containing papain, are stabilized by the addition of salt, propylene glycol, and glucose. The water activity of these preparations is not known, but it is probably in the

0.6–0.8 range. A similar stabilizing effect on a bacterial protease by polyols (sorbitol, propylene glycol, and glycerol) was observed by Yasumatsu *et al.* (1965) and is described earlier in this chapter. It should also be noted that the stabilization of liquid enzyme preparations by the addition of high (40–80%) concentrations of glycerol has long been a standard practice in enzymology.

In one of the few kinetic studies of trypsin activity at reduced moisture levels, Inagami and Sturtevant (1960) investigated the hydrolysis of the ethyl ester of benzoyl-arginine in dioxane–water mixtures and found that the addition of 50% of the humectant dioxane to the reaction mixture increased the reaction rate. Beyond this concentration, trypsin activity was reduced, but at 88% dioxane levels, 68% of the maximal reaction rate remained. Although the water activity levels of the test systems were not measured, it is apparent that this enzyme functions in systems containing relatively small amounts of free water.

Invertase

Invertases, enzymes which convert sucrose to glucose and fructose, are used extensively in the confectionery and syrup industries. The ability of this enzyme to liberate a sugar of greater solubility and sweetness (fructose) is the principal reason for its wide application. The basic chemistry of the action of invertase rests on the firm kinetic basis established in 1913 by Michaelis and Menten, who theorized that the rate-limiting step of enzyme reactions is the formation of an enzyme–substrate complex. The hydrolysis of sucrose by invertase was found subsequently to confirm the initial, theoretical considerations, and it is at least partly on these studies that the science of enzymology is now based. Michaelis and Menten studied the hydrolysis of solutions of cane sugar in the range of 0.007–0.385 M (0.999–0.986 a_w), and found maximal velocities occurring at 0.192 M (0.993 a_w). Others (Nelson and Schubert, 1928) observed that the velocity of the reaction decreased when the sucrose concentration was greater than 10% (0.994 a_w) and concluded that the decrease in water concentration was the critical factor in the drop in velocity of the inversion reaction. For a time it was believed that increase in viscosity (decrease in fluidity) of these solutions is responsible for this decrease in invertase activity. However, additional experiments, in which sucrose concentrations were maintained constant while the water content was varied by the addition of various solutes, such as glycerol, pectin, or ethyl alcohol, established that water content, rather than viscosity, is the critical factor. The demonstration that moisture content limited invertase activity was eventually extended to food systems by Kertesz (1935), who showed that hydrolysis of sucrose by invertase in apple pulp occurs at 4.75% water, but not at 4%. From these and other studies, it can be concluded that optimal a_w levels for invertase activity are greater than 0.997. Minimal a_w levels have not been well defined up to the present time. As in the case of other enzyme systems, enzyme purity, substrate composition, temperature, pH, and a host of other factors may alter markedly the minimal and optimal a_w levels.

Lipoxidase, Peroxidase, Phenoxidase

Lipoxidase finds its principal commercial application in the baking industry, where this enzyme, usually obtained by extraction of soy flours, is used as a dough whitening or bleaching agent. Conversely, in the production of certain flours, e.g., durum varieties, a yellow color is desirable, and in pasta doughs, which utilize high concentrations of durum flour, lipoxidase activity must be suppressed. Alternatively, flours low in lipoxidase content may be used. The manipulation of the activity of this enzyme through a_w alterations, although not currently employed commercially, could be a practical proposition. Blaine (1962) investigated the effect of glycerol/water mixtures of varying composition on the reaction velocities of peroxidase and lipoxidase (Fig. 3.5), as noted earlier. Both enzymes appear to respond to limited moisture conditions in the manner noted previously in this chapter (see section on Model Systems). There appears to be a slight decline in peroxidase activities at a_w levels in excess of 0.97, an effect which does not occur with lipoxidase. Both enzymes exhibit optimal activity in the range of 0.94–0.97 a_w. Glycerol and sugars generally tend to have less inhibitory effect on peroxidase activity than organic solvents such as dioxane or ethyl alcohol.

Phenoxidases, or polyphenol oxidases, are commonly found in various plants and are responsible for the rapid discoloration of many fruits and vegetables subsequent to slicing or bruising. This enzyme appears to be extremely sensitive to NaCl; however, the low concentrations (0.1–0.2%) at which this inhibition is manifest indicate that inhibition by this salt is intrinsic, rather than a consequence of a decrease in a_w. Optimal and minimal a_w levels for the activity of phenoxidases have not been determined in model systems or in foods using solutes other than NaCl to adjust moisture conditions. In his extensive review of enzymatic reactions in low-moisture foods, Acker (1962) cites a number of findings indicating that this class of enzymes is relatively inactive in dried food products in the 5–12% moisture range.

Amylases

Amylases, both α- and β-, are used principally in the production of corn syrups and in the baking industry. In addition, sizable quantities of these enzymes are utilized in the production of fermentable carbohydrates from cereal starches for the brewing and alcoholic spirits industries.

Most studies of this group of enzymes have been on model systems, consisting usually of starch/water mixtures and crude or partially purified enzymes. Kier-

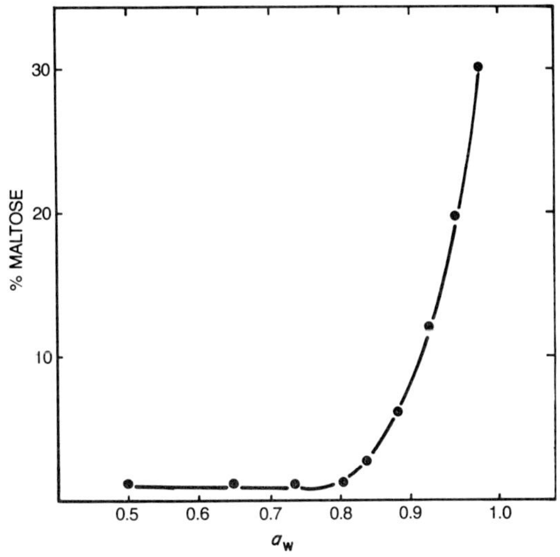

Fig. 3.6. Effect of a_w on β-amylase activity (Drapron, 1972).

meier and Corduro (1954) hydrated starch–enzyme mixtures to various water contents and found that enzymatic activity increased with increasing percent water, with a maximum occurring at about 42%. On the other hand, amylase activity could not be demonstrated below 14% water. If, however, these reactions were carried out on a porous but inert matrix (in this case, a cellulose filter pad), the minimal level at which the reaction would occur was reduced to < 10% moisture. This phenomenon was attributed to the increase of free water on the capillary surfaces of the matrix.

Drapron (1972) used maltose formation from starch as a measure of β-amylase activity and found a sharp inflection point at 0.80–0.83 a_w. Beyond this region, enzyme activity increased dramatically, whereas from 0.50 (the lowest test value) to 0.80 a_w, there was little or no increase in activity (Fig. 3.6). If, instead of hydrating the starch to given a_w levels, this author adjusted the a_w of his test systems with added glycerol or lithium chloride, much greater reactivity was observed at given a_w levels. Two possible explanations exist for these results: intrinsic inhibition by the solutes or changes in the capillarity of the starch/water system as discussed above. Further investigations by Drapron have shown that the response of β-amylase activity to water content is relatively temperature independent in the range 21°–31°C, and that enzyme reaction rates, as well as total amount of product (maltose) formed, increase with increasing water activity.

The water relations of starch hydrolysis by α-amylase may be somewhat different from those of β-amylase. There is some evidence (Drapron and Guilbot, 1962) that the products of starch hydrolysis are drastically affected by the moisture content of the system. In the "dry state," or at low a_w levels, only maltose and glucose appear and oligosaccharides are not formed until the a_w of these systems is raised. These effects, especially if found to occur in other enzyme systems, could be of great significance in the applications of enzymes in the food industry and, at the very least, deserve additional study. In such investigations, both substrate and reaction products are solutes that may be present in high concentrations and, hence, capable of significantly influencing the water activity. Thus, great care must be taken to ensure that the reactions are occurring at specified a_w levels, and that a_w measurements prior to, during, and following the reaction period are made. As pointed out earlier, enzyme purity must also be established.

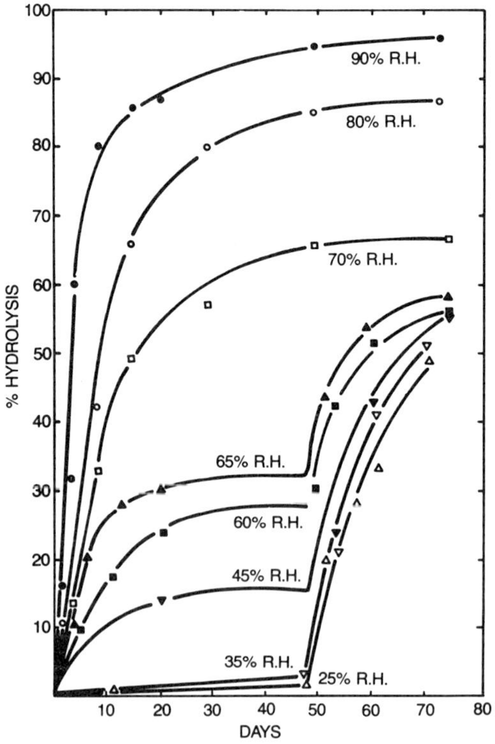

Fig. 3.7. Effect of relative humidity on lecithin hydrolysis. Water activity adjusted to 70% R.H. at 48 hours in 25, 35, 45, 60, and 65% R.H. systems (Acker and Kaiser, 1961).

Phospholipase or Lecithinase

The principal studies of the effect of relative humidity on the activity of phospholipases are those of Acker and Luck (1958, 1959) and Acker and Kaiser (1961). Only minimal hydrolysis occurs at equilibrium relative humidity levels below 35% (Fig. 3.7). Reaction velocity continues to increase through 90% relative humidity, the highest level tested. After 48 days of incubation, these authors repoised the relative humidities of samples held from 25 to 65% to a humidity level of 70%. An immediate resumption of hydrolytic activity followed. Thus, the inhibition of this enzyme by reduced a_w levels does not involve permanent damage to the enzyme, or, at least, to its reactivity.

There are many additional enzymes that find application in foods and which would be suitable subjects for investigations of their water relations. Unfortunately, few detailed studies of this type have been reported, and we must rely on indirect evidence provided by studies in dry or low moisture foods. The thorough review by Acker (1962) provides numerous references of this type.

Obviously, much work remains in order to define the optimal and minimal a_w levels for the action of food enzymes. Model systems studies are required, utilizing enzymes of defined purity under closely controlled conditions. In studies of this type, reaction rates should be calculated, not only on product formation, but also on substrate depletion. Once determined, reaction processes should be mechanistically and kinetically evaluated in keeping with established thermodynamic principles. Special emphasis should be given to the possibility that not only reaction rates, but also reaction products, can vary as a function of water activity alterations.

EFFECT OF WATER ACTIVITY ON NONENZYMATIC BROWNING REACTIONS

Nonenzymatic browning occurs via a series of complex, defined reactions in which the initial reactants are usually reducing sugars and the amino groups of amino acids or proteins. Under suitable conditions (temperature, pH, water activity) these substrates proceed via the formation of a number of intermediates to highly colored, heterocyclic nitrogen compounds of varying composition. The nature of these reactions and their relationship to a proposed scheme of type reactions was reviewed in 1953 by Hodge, who emphasized the significance of the formation of amino ketoses via Amadori rearrangement as a key reaction in the formation of browning or Maillard reaction products. The subsequent reviews of Reynolds (1963, 1965), Peer (1975), and Shallenberger and Birch (1975) describe the great deal of work on the chemistry of nonenzymatic browning after 1953.

Much of what is now known of the mechanisms by which browning reactions occur and of the effect of water activity on these reactions has been derived from earlier studies on model systems. In one of the best of these, Lea and Hannan (1949) measured the rate of loss of free amino nitrogen (commensurate with increase in browning reaction) from casein–glucose solutions adjusted to various a_w levels and maintained at various temperatures. Maximal loss of amino nitrogen occurred in samples equilibrated to 0.65–0.70 a_w. This loss was reduced at higher and lower humidities. The optimal a_w was the same at temperatures of 37°, 70°, and 90°C. The increase in rate of browning as the a_w approaches 0.65–0.70 can best be explained by reference to the sorption isotherm. As the moisture content of the system increases, the isotherm becomes virtually linear. This linear increase is believed to indicate the establishment of a double layer of water molecules between protein laminates, with the result that exposed protein polar groups become saturated with adsorbed water. Protein molecules increase greatly in mobility, and the possibility of intra- and intermolecular rearrangement is thereby enhanced, permitting the observed increase in the browning reaction. If the a_w of these systems is increased beyond 0.70, as noted earlier, the reaction rate once again decreases because of the mass action effect when simple dilution reduces the quantity of substrate (i.e., glucose) available for reaction (Shallenberger and Birch, 1975). Another hypothesis proposed by Lea and Hannan (1949) states that an increasing "thickness" of water layers or "aqueous film" could separate amino and potential aldehyde groups, thus preventing their interaction.

This possible mechanism for the effect of increasing a_w levels on the browning reaction, while based in part on theoretical considerations, explains many of the experimental data obtained since Lea and Hannan first proposed it in 1949. As more sophisticated analytical techniques, such as nuclear magnetic resonance and electron spin resonance, have been applied to studies on the nature of the browning reaction and the influence of water activity and water binding upon it, the basic mechanism outlined above has received substantial support.

Model browning systems have been studied by other workers (Tannenbaum, 1966; Reynolds, 1963, 1965; Schwartz and Lea, 1952), but few have chosen to investigate water activity as an experimental variable. Many of the reports that have appeared have emphasized the difficulty of predicting the optimal rates of browning wholly on a_w considerations. In retrospect, this is not surprising considering the dependence, at least in theory, of this reaction on the conformation of the sorption isotherm and the almost infinite variations in isotherms existing in foods. In addition, variability in protein water-binding sites and planar configuration have further complicated the issue. Because of these complex interactions, a_w optima for browning of 0.3–0.8 are not unexpected. Although a_w is clearly important in browning reactions, other factors, such as pH, temperature, and viscosity (Eichner and Karel, 1972), also may influence markedly the progress of these reactions.

As stated earlier, the substrate requirements for the browning reaction are essentially a reducing sugar and a reactive amino group. Frequently, however, browning occurs in the absence of reducing sugars in systems containing initially, e.g., sucrose, as the only sugar. In such systems, it is the hydrolysis of sucrose to the reducing sugars, glucose and fructose, that may become the rate-limiting step in browning. Schoebel *et al.* (1969) investigated sucrose hydrolysis at limited moisture levels and found that, in unsaturated solutions, the rate of sucrose dissolution becomes limiting when the initial amounts of dissolved sucrose have been consumed. Although these authors did not directly relate water activity to the progress of the hydrolytic reaction, the extent of sucrose solubility and the effect of the invert sugars produced on this solubility appeared to be the key factors affecting hydrolysis.

Browning Reactions in Foods

Browning reactions that occur during food processing or storage may either increase or decrease the acceptability of the food. In either case, the enlightened manipulation of water activity, as discussed in the section on model systems, can do much to contribute to the acceptability of a food.

The browning of a typical food system corresponds generally to that of model systems. Usually, the rate and extent of the reaction increase with increasing a_w until a maximal level is reached, usually at 0.60–0.80 a_w, followed by a decrease in browning rates with further increases in excess of 0.75–0.85 a_w. A typical

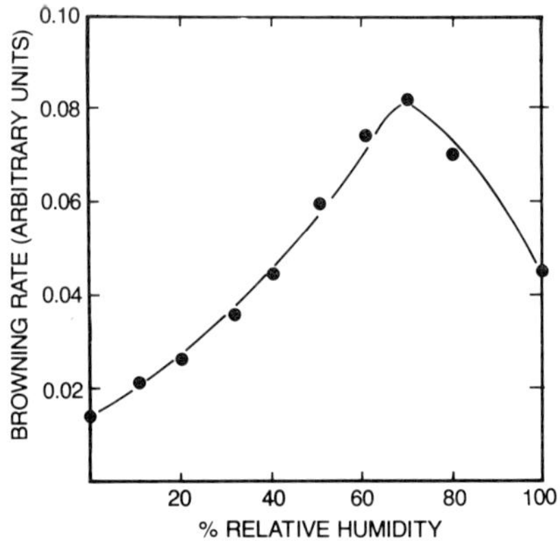

Fig. 3.8. Effect of relative humidity on the browning rate of dehydrated pea soup mix at 54°C (Labuza *et al.*, 1970).

browning rate curve as related to percent relative humidity ($a_w \times 100$) is shown in Fig. 3.8. In this figure, a steady increase in rate occurs as the relative humidity of the system (dehydrated pea soup) is increased, with a maximal rate occurring in the 65–70% R.H. range. Induction time for the reaction also appears to be affected by the moisture content of the food. Labuza *et al.* (1970) noted that at 54°C, dehydrated skim milk samples held at 32, 50, and 75% R.H. had extrapolated browning induction times of 16, 2.5, and 1 days, respectively. In practice, the induction times for browning of dehydrated foods may be extended because of the extremely low a_w of these foods, and long-term storage studies of a year or more may be necessary to evaluate properly the susceptibility of a food product to browning. Mizrahi *et al.* (1970) investigated the feasibility of using accelerated storage tests with dehydrated cabbage by increasing, separately or together, the temperature of storage and the moisture content of the food. The activation energy for browning was dependent on moisture content, and mathematical formulae were developed which allowed extrapolation from high- to low-moisture foods with a reduction in storage time from greater than 1 year to approximately 10 days. Additional studies exist which relate to the effect of moisture on nonenzymatic browning of dehydrated potatoes (Hendel *et al.*, 1955), dried orange juice (Karel and Nickerson, 1964), and dehydrated lemonade and pineapple juice (Notter *et al.*, 1955, 1958).

The extent of browning reactions in meat and fish products, especially during drying and storage, is of great importance. In addition to deleterious color changes, browning reactions in meat products usually result in bitter or burned flavor notes. Sharp (1962) investigated the influence of E.R.H. on the browning deterioration of precooked freeze-dried pork. At 37° and 50°C, browning reaction increased as relative humidity rose to 57%, followed by a decrease as it was further increased to 70%. Flavor deterioration in all of the samples occurred over the entire storage period (477 days). At the conclusion of the experiment, only those samples equilibrated to 16% R.H. were judged to be "acceptable."

The browning reaction which occurs in dried cod (and presumably in other fish muscle products) is principally the result of a Maillard-type reaction between amino groups and ribose (Jones, 1954, 1956). Browning reaction rates in these systems (Fig. 3.9) are maximal at much lower relative humidity levels than in vegetable products or in the model systems tested by Lea and Hannan (1949). Thus, extremely low relative humidity levels (in the range of 5–10%) seem to be required to stabilize the color of cod muscle. Although significant resistance to browning occurs at relative humidities > 75%, the susceptibility of cod and other foods to microbial growth and spoilage at these levels would, under most circumstances, preclude the use of this method of control. Minimal browning of dried anchovies has been reported by Han *et al.* (1973) at moisture levels (7–9%) similar to those for cod. Corresponding a_w levels for anchovy flesh at 7–9% moisture are 0.32–0.45.

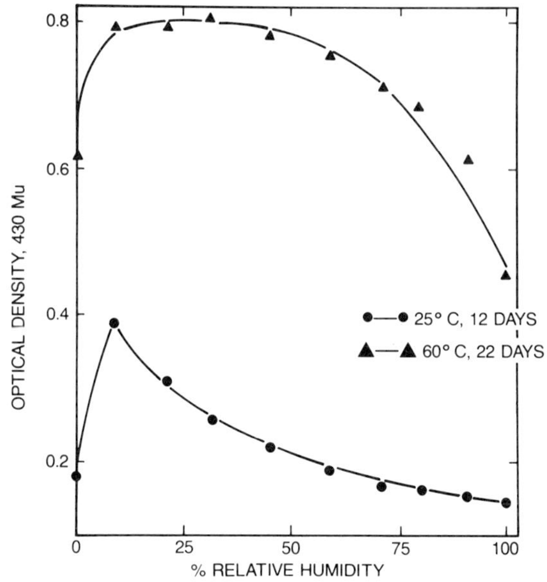

Fig. 3.9. Browning of cod muscle at various relative humidity levels (Jones, 1954).

Nonenzymatic browning is a serious problem in the production of intermediate moisture foods (IMF). Such foods are normally poised at 0.6–0.8 a_w (see Chapter 9), which places them well within a_w ranges for optimal browning. In addition to deleterious color reactions, such foods also may undergo losses in nutrient quality as a result of reactions involving proteins and amino acids, which form Maillard browning products.

The possibility of reducing the extent of browning in IMF by the use of glycerol as a humectant has been suggested by the work of Warmbier *et al.* (1976). These workers found that the a_w at which maximal browning occurs was reduced from the 0.75 to 0.65 a_w range to 0.50–0.40 a_w when glycerol was used to adjust the a_w of a foodlike model system. This downward shift was attributed to the liquid, waterlike properties of glycerol, properties which maintain reactant mobility and, thereby, reactivity at low a_w levels.

Much additional work remains before the initiation and extent of browning in foods as a function of a_w can be predicted and ultimately controlled. Present data suggest that maximal browning reaction rates in vegetable and fruit products occur in the 0.65–0.75 a_w range. The less complete data available on meat and muscle products show a broader optimal a_w range for browning, usually in the range of 0.30–0.60 a_w. Dried dairy products, principally nonfat dry milk, appear to brown most readily at approximately 0.70 a_w. Variations, both chemical and physical, in individual food products and differences in the ways that foods are processed make exact a_w–browning predictions difficult. Despite these problems, the adjustment of water activity remains one of the most promising techniques for controlling the browning of many foods.

REFERENCES

Acker, L. (1962). Enzymic reactions in foods of low moisture content. *Adv. Food Res.* **11,** 263–330.

Acker, L. (1963). Enzyme activity at low water contents. *Recent Adv. Food Sci.* **3,** 239–247.

Acker, L., and Beutler, H. O. (1963). Die enzymatische Phytinspaltung in geschrotetem Getreide and Abhängigkeit von der relativen Luftfeuchtigkeit. *Z. Ernaehrungswiss., Suppl.* **3,** 1–5.

Acker, L., and Kaiser, H. (1959). Uber den Einfluss der Feuchtigkeit auf den Ablauf enzymatischer Reaktionen in wasserarmen Lebensmitteln. II. *Z. Lebensm.-Unter.-Forsch.* **110,** 349–356.

Acker, L., and Kaiser, H. (1961). Uber den Einfluss der Feuchtigkeit auf den Ablauf enzymatischer Reaktionen in wasserarmen Lebensmitteln. III. *Z. Lebensm-Unters.-Forsch.* **115,** 201–210.

Acker, L., and Luck, E. (1958). Über den Einfluss der Feuchtigkeit auf den Ablauf enzymatischer Reaktionen in wasserarmen Lebensmitteln. *Z. Lebensm.-Unter.-Forsch.* **108,** 256–269.

Acker, L., and Luck, E. (1959). Uber das Vorkommen von Enzymen, in besondere Phospholipasen, in Eieren und Eitrockenerzeugnissen. *Dtsch. Lebensm.-Rundsch.* **55,** 242–246.

Blaine, J. A. (1962). Moisture levels and enzyme activity. *Recent Adv. Food. Sci.* **2,** 41–45.

Caillot, J. M., and Drapron, R. (1970). Unpublished results, cited in Drapron (1972).

Cuendet, L. S., Larson, E., Norris, C. G., and Geddes, W. F. (1954). The influence of moisture content and other factors on the stability of wheat flours at 37.8°C. *Cereal Chem.* **31,** 362–389.

Drapron, R. (1972). Reactions enzymatiques en milieu peu hydraté. *Ann. Technol. Agric.* **21,** 487–499.

Drapron, R., and Guilbot, A. (1962). Contribution à l'étude des réactions enzymatiques dans les milieux biologiques peu hydratés. La dégradation de l'amidon par les amylases en fonction de l'activité de l'eau et de la température. *Ann. Technol. Agric.* **11,** 175–218.

Eichner, K., and Karel, M. (1972). The influence of water content and water activity on the sugar-amino browning reaction in model systems under various conditions. *J. Agric. Food Chem.* **20,** 218–223.

Guardia, E. J., and Haas, G. J. (1967). Influence of water binders on the activity and thermal inactivation of lipase. *J. Agric. Food Chem.* **15,** 412–416.

Han, S. B., Lee, J. H., and Lee, K. H. (1973). Non-enzymatic browning reactions in dried anchovy when stored at different water activities. *Bull. Korean Fish. Soc.* **6,** 37–43.

Hendel, C. E., Silveira, V. G., and Harrington, W. O. (1955). Rates of non-enzymatic browning of white potato during dehydration. *Food Technol. (Chicago)* **9,** 433–438.

Hodge, J. E. (1953). Dehydrated foods. Chemistry of browning reactions in model systems. *J. Agric. Food Chem.* **1,** 928–943.

Inagami, T., and Sturtevant, J. M. (1960). Trypsin-catalyzed hydrolysis of benzoyl-1-arginine ethyl ester. I. Kinetics in dioxane-water mixtures. *Biochim. Biophys. Acta* **38,** 64–79.

Jencks, W. P. (1969). "Catalysis in Chemistry and Enzymology." McGraw-Hill, New York.

Jones, N. R. (1954). 'Browning' reactions in freeze-dried extractives from the skeletal muscle of codling (*Gadus callarias*). *Nature (London)* **174,** 605–606.

Jones, N. R. (1956). Discoloration of muscle preparations from codling (*Gadus callarias*) by degradation products of 1-methylhistidine. *Nature (London)* **177,** 748–749.

Karel, M., and Nickerson, J. T. R. (1964). Effects of relative humidity, air and vacuum on browning of dehydrated orange juice. *Food Technol. (Chicago)* **18,** 1214–1218.

Kertesz, Z. I. (1935). Water relations of enzymes. II. Water concentration required for invertase action. *J. Am. Chem. Soc.* **57,** 1277–1279.

Kiermeier, F., and Corduro, E. (1954). Uber den diastatischen Starkeabbau in lufttrockenen Substanzen. *Biochem. Z.* **325,** 280–287.

Labuza, T. P., Tannenbaum, S. R.,and Karel, M. (1970). Water content and stability of low-moisture and intermediate-moisture foods. *Food Technol. (Chicago)* **24,** 543–550.

Lea, C. H., and Hannan, R. S. (1949). Studies of the reaction between proteins and reducing sugars in the "dry" state. I. The effect of activity of water, of pH and of temperature on the primary reaction between casein and glucose. *Biochim. Biophys. Acta* **3,** 313–325.

Matz, S., McWilliams, C. S., Larsen, R. A., Mitchell, J. H. McMullen, J., and Layman, B. (1955). The effect of moisture content on the storage deterioration rate of cake mixes. *Food Technol. (Chicago)* **9,** 276–285.

Michaelis, L., and Menten, M. L. (1913). Kinetics of invertase action. *Biochem. Z.* **49,** 333–369.

Mizrahi, S., Labuza, T. P., and Karel, M. (1970). Feasibility of accelerated tests for browning in dehydrated cabbage. *J. Food Sci.* **35,** 804–807.

Nelson, J. M., and Schubert, M. P. (1928). Water concentration and the rate of hydrolysis of sucrose in invertase. *J. Am. Chem. Soc.* **50,** 2188–2193.

Notter, G. K., Taylor, D. G., and Walker, L. H. (1955). Stabilized lemonade powder. *Food Technol. (Chicago)* **9,** 503–505.

Notter, G. K., Taylor, D. H., and Brekke, J. E. (1958). Pineapple juice powder. *Food Technol. (Chicago)* **12,** 363–366.

Peer, H. G. (1975). Degradation of sugars and their reactions with amino acids. *Landbouwhogesch. Wageningen, Misc. Pap.* Part 9, pp. 109–115.

Poisson, J., and Guilbot, A. (1963). Storage conditions and the length of grain storage. *Meun. Fr.* **193,** 19–29.

Purr, A. (1966). Zum Ablauf chemischer Veränderungen in wasserarmen Lebensmitteln. I. Der Enzymatische Abbau der Fette bei niedrigen Wasserdampf-partialdrücken. *Fette, Seifen, Anstrichm.* **68,** 145–154.

Reed, G. (1966). "Enzymes in Food Processing," 1st ed. Academic Press, New York.

Reynolds, T. M. (1963). Chemistry of nonenzymic browning. I. *Adv. Food Res.* **12,** 1–52.

Reynolds, T. M. (1965). Chemistry of nonenzymic browning II. *Adv. Food Res.* **14,** 167–283.

Rockland, L. B. (1957). A new treatment of hygroscopic equlibria: Application to walnuts (*Juglans regia*) and other foods. *Food Res.* **22,** 604–628.

Rothe, M. (1958). Eigenschaften and Reaktions verhalten pfanzlicher Samenlipasen im natürlichen Mileu. I. Mitt. Aktivitätsbestimmung unter natürlichen Milieubedingungen. *Ernaehrungsforschung* **3,** 21–40.

Sapers, G. M., and Nickerson, J. T. R. (1962). Stability of spinach catalase. III. Regeneration. *J. Food Sci.* **27,** 287–290.

Schoebel, T., Tannenbaum, S. R., and Labuza, T. P. (1969). Reaction at limited water concentration. I. Sucrose hydrolysis. *J. Food Sci.* **34,** 324–329.

Schwartz, H. M., and Lea, C. H. (1952). The reaction between proteins and reducing sugars in the "dry" state. Relative reactivity of the α- and ϵ-amino groups of insulin. *Biochem. J.* **50,** 713–716.

Shallenberger, R. S., and Birch, G. G. (1975). "Sugar Chemistry." Avi Publ. Co., Westport, Connecticut.

Sharp, J. G. (1962). Nonenzymatic browning deterioration in dehydrated meat. *Recent Adv. Food Sci.* **2,** 65–73.

Snow, D., Crichton, H. G., and Wright, N. C. (1944). Mould deterioration of feeding-stuffs in relation to humidity of storage. Part II. The water uptake of feeding-stuffs at different humidities. *Ann. Appl. Biol.* **21,** 111–116.

Tannenbaum, S. R. (1966). Protein carbonyl browning systems: A study of the reaction between glucose and insulin. *J. Food Sci.* **31,** 53–57.

Warmbier, H. C., Schnickels, R. A., and Labuza, T. P. (1976). Effect of glycerol on nonenzymatic browning in a solid intermediate moisture food system. *J. Food Sci.* **41,** 528–531.

Yasumatsu, K., Ono, M., Matsumura, C., and Shimazono, H. (1965). Stabilities of enzymes in polyhydric alcohols. *Agric. Biol. Chem.* **29,** 665–671.

4

Lipid Oxidation, Changes in Texture, Color, and Nutritional Quality

EFFECT OF a_w ON LIPID OXIDATION

Lipid oxidation, or oxidative rancidity of fat, as it is frequently termed, is a type of autoxidation that occurs when olefinic compounds are oxidized by atmospheric oxygen. The result of this oxidative process in foods is the formation of off-flavors or flavor reversions in lipids or lipid-containing foods, the destruction of fatty acids (including those considered essential in human diets), and the destruction of certain vitamins. In most cases, these changes are deleterious, but some foods obtain their distinctive flavors from either hydrolytic rancidity (enzyme-mediated formation of free fatty acids) or by oxidative rancidity.

Free-radical-producing substances and other factors, such as transition metal ions and light, may catalyze lipid oxidation. This process normally occurs as a complex series of reactions involving the formation of hydroperoxides, usually during an induction period. This period is followed by a phase in which the generation of free radicals is self-propagating. A third phase, called termination, may take place,in which the various hydroperoxy compounds and free radicals react with each other to form stable products.

During the induction phase, a hydrogen atom is removed from the carbon atom adjacent to the double bond of an unsaturated fatty acid [Eq. (1)]:

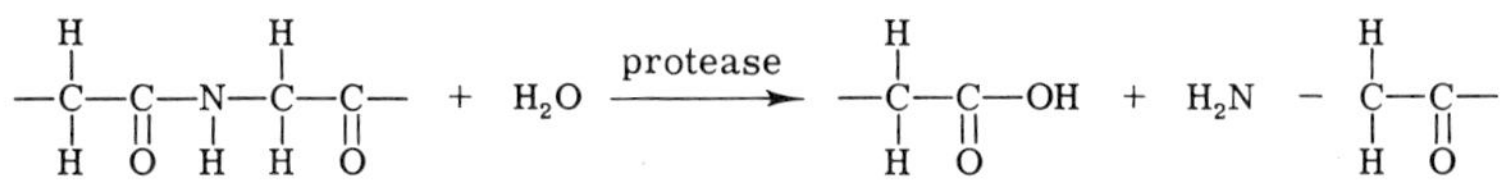

(1)

Trace quantities of transition metals, light, or enzymes can catalyze this reaction. The free radical (termed a resonating radical) may react with oxygen to produce peroxyl radicals [Eq. (2)]:

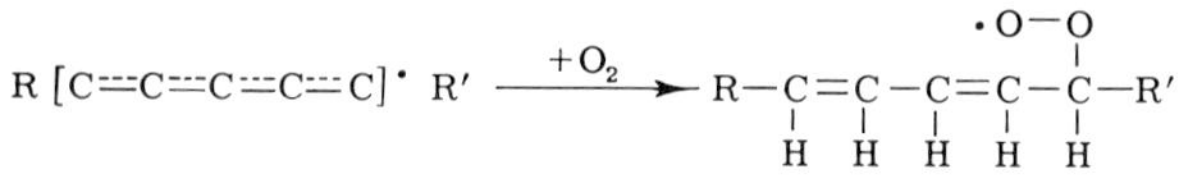

$$R [C\!=\!C\!=\!C\!=\!C\!=\!C]^{\cdot} R' \xrightarrow{+O_2} R-\underset{H}{\overset{\overset{\displaystyle \cdot O-O}{|}}{C}}=\underset{H}{\overset{}{C}}-\underset{H}{\overset{}{C}}=\underset{H}{\overset{}{C}}-\underset{H}{\overset{}{C}}-R' \qquad (2)$$

peroxyl radical

Peroxyl radicals are then reduced to form hydroperoxides, the hydrogen ion coming from other unsaturated fatty acid molecules, thus creating more free radicals and so propagating the reaction [Eq. (3)].

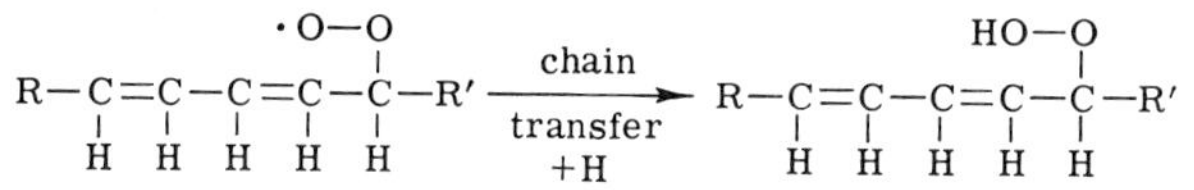

$$R-\underset{H}{\overset{\overset{\displaystyle \cdot O-O}{|}}{C}}=\underset{H}{\overset{}{C}}-\underset{H}{\overset{}{C}}=\underset{H}{\overset{}{C}}-\underset{H}{\overset{}{C}}-R' \xrightarrow[+H]{\text{chain transfer}} R-\underset{H}{\overset{}{C}}=\underset{H}{\overset{}{C}}-\underset{H}{\overset{}{C}}=\underset{H}{\overset{}{C}}-\underset{H}{\overset{\overset{\displaystyle HO-O}{|}}{C}}-R' \qquad (3)$$

hydroperoxide product

Scission of these products can then occur, with resultant production of aldehydes, ketones, and short-chain fatty acids, which ultimately are responsible for rancid flavors and off-odors.

In addition to metals (especially iron and copper), a number of other factors may influence autoxidation. The nature of the lipid itself is extremely important. A lipid containing a high proportion of unsaturated fatty acids is much more susceptible to oxidative rancidity development than a less saturated fat. Temperature also has a marked effect on autoxidation, principally in relation to propagation [Eq. (3)] and to the decomposition of alkyl peroxides. Under most circumstances the rate of autoxidation increases with temperature, so that foods susceptible to oxidative rancidity that are stored at reduced temperatures, are more stable.

From Eq. 2, it can be seen that molecular oxygen is required for autoxidation. Conversely, the removal of oxygen, as in vacuum-packaged foods, or its replacement by an inert gas, as in nitrogen packing, will retard or prevent the formation of hydroperoxy radicals and will thereby interfere with the oxidative process.

As mentioned earlier, light also will accelerate these processes. Thus, foods susceptible to oxidative deterioration often are packaged in containers impervious to certain wavelengths of the spectrum. Ionizing radiation also renders foods more susceptible to autoxidation, a factor that has contributed to the delay in the development and acceptance of food preserved by this means.

The effect of moisture on the development of oxidative rancidity in foods differs markedly from its effect on other food reactions, such as nonenzymatic browning, enzymatic activity, and textural changes. At very low water activity levels, foods containing unsaturated fats and exposed to atmospheric oxygen are highly susceptible to the development of oxidative rancidity. This high oxidative activity occurs at water activity levels below the so-called monolayer level of moisture. As water activity is increased, both the rate and the extent of autoxidation decrease until an a_w in the range 0.3–0.5 is reached, depending upon the system investigated. At this point, the rate of oxidation increases and continues to do so through the intermediate moisture food (IMF) range until a steady state is reached, normally at a_w levels in excess of 0.75. Although the relative rate of autoxidation has been shown to decrease in model systems at still higher a_w levels (Labuza, 1971), the present authors are aware of no evidence to support this suggestion in foods.

Mechanisms

Although there is little disagreement on the overall effect of moisture content on oxidative rancidity, controversy remains on the mechanisms by which these reactions occur. Because foods vary greatly in their content of the factors essential for the onset of oxidative rancidity, it is difficult to predict exactly how a specific food will react under all circumstances of production and storage. Thus, a thorough understanding of the mechanisms involved and the interaction of these mechanisms is basic to obtaining acceptable and wholesome foods.

To obtain a degree of familiarity with these principles, one may interpret the effects of moisture in terms of the monolayer water absorption theory described in Chapter 1. Salwin (1959) regarded this monolayer as a protective film that interferes with autoxidation by preventing oxygen from reacting with unsaturated lipids. Thermodynamically, this interference cannot represent a "blanket" of water molecules forming a continuous film of water, but rather, may correspond to the quantity of hydrogen-bonding sites on the food molecules. The "protective" effect exerted by water creates bond energy spheres that inhibit reactions with adjacent polar groups and thereby contributes other desirable attributes to the food, such as increased hydratability and improved texture.

Others have attempted to explain the high rates of oxidation at very low a_w levels in terms of free radical reactions occurring at or below the BET monolayer. Using electron paramagnetic resonance (EPR) techniques to study the fate of "trapped" free radicals in oxidizing lipid–protein systems, Roubal (1970) found that free radicals readily react with fish proteins to produce excessive polymerization and amino acid destruction in systems low in moisture. Munday *et al.* (1962) detected electron-spin resonance signals characteristic of free radicals in freeze-dried foods, but not in other dried foods that were not freeze-dried. Radical concentrations increased in samples containing high levels of lipids when the temperature was increased to 100°C in the presence of air and

O_2 but declined in the presence of added water. The decrease in free radical signals to undetectable levels with the increase in moisture was attributed to the ability of water molecules to dislodge oxygen from the dried tissues.

Many of the hypotheses concerning the mechanism of the effect of moisture on autoxidation involve foods which, by their very nature, are extremely complex systems. In addition, the presence of interfering reactions often makes the evaluation of experimental data difficult. It was with this shortcoming in mind that investigators at the Massachusetts Institute of Technology undertook a series of model system studies utilizing methyl linoleate and fatty acids as substrates to study free radical formation and its relationship to autoxidation at various moisture levels. The first of these reports (Maloney *et al.*, 1966) described the use of oxygen uptake as a measure of oxidation and described a definite retardation in uptake as the a_w of the model system was increased. The authors related this effect to the hydrogen bonding of hydroperoxides to water, with the result that hydroperoxide decomposition is prevented. Less hydroperoxide would thus be available for initiation of the oxidation process.

In similar studies on cobalt-catalyzed autoxidation of methyl linoleate, Labuza *et al.* (1966) found that increases in water activity are inhibitory in the presence or absence of cobalt salts. In all cases, humidification (to 29% relative humidity) substantially increased the time required for free radical initiation to occur. It was suggested that somewhat higher concentrations of hydroperoxides are required to initiate the bimolecular phase of induction in the "high" humidity systems. These authors interpreted their data in terms of the affinity of the water–lipid interface for the amphipolar hydroperoxides where these hydroperoxides are "trapped" due to the formation of hydrogen bonds. The result of this "entrapment" is the observed extension of the oxidation induction period. In addition to this mechanism, it was suggested that the addition of water to methyl linoleate hydrates metal catalysts by hydration of coordination shells, thus preventing the catalysts from acting as free radical inducers. In a subsequent publication (Labuza *et al.*, 1969), the effectiveness of antioxidants at various humidities was also described (to be dealt with later in this chapter), and oxidation was shown to become increasingly rapid at much higher humidity levels (> 40% R.H.). A catalytic effect observed as a result of moisture addition in this range and above was attributed to the greater mobility of reactants. However, the situation in foods at high a_w levels may be somewhat different. In such cases, the hydration to a_w levels above 0.40 can cause physical swelling of proteins or carbohydrates and the exposure of additional oxidizable sites. This would of course increase oxidation rates, as was, in fact, observed in systems of this type.

In studies described thus far, relative humidity adjustments were obtained by freeze-drying the model system (usually on a cellulose carrier) and hydrating it to a specified a_w level by equilibration. Thus, specific effects caused by a_w adjustment with solutes could be avoided. In practice, however, foods may be adjusted

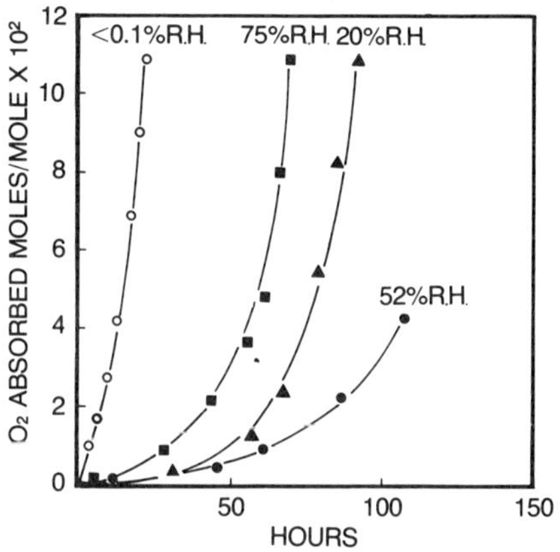

Fig. 4.1. Effect of relative humidity on the uptake of oxygen by a cellulose-containing model system without glycerol (Heidelbaugh and Karel, 1970).

to specific a_w levels by additions of solutes, such as NaCl or glycerol. Heidelbaugh and Karel (1970) investigated the effect of a number of solutes on the oxidation of a model system containing methyl linoleate and ground pork. In the case of glycerol, which displaces water from the oxidizable substrate, the transition point (the point at which further water addition changes the net reaction from antioxidant to prooxidant effects) is shifted from 0.52 a_w in the control system to 0.11 a_w in the glycerol-containing system. This decrease, shown in Figs. 4.1 and 4.2, was attributed to the binding of water to glycerol and the participation of this bound water in oxidation reactions. Thus, the prooxidant effect of the

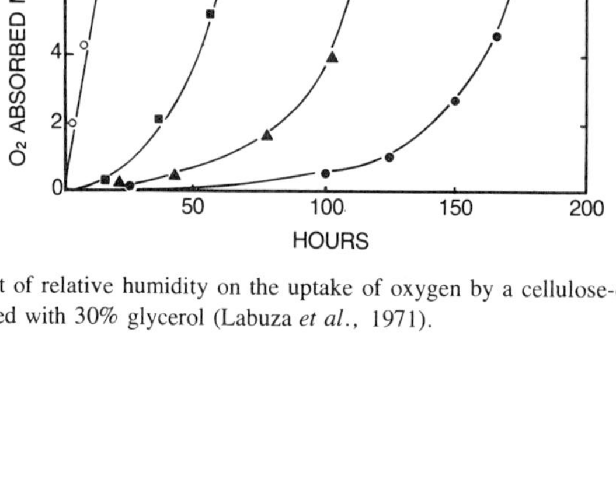

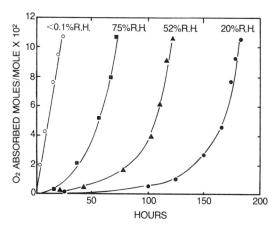

Fig. 4.2. Effect of relative humidity on the uptake of oxygen by a cellulose-containing model system supplemented with 30% glycerol (Labuza *et al.*, 1971).

glycerol–water combination persists in all but extremely low a_w systems. The opposite effect (an increase in the transition a_w level) was noted by Heidelbaugh *et al.* (1971) when dextran was added to a methyl linoleate-containing system. Thus, the type of solute controlling a_w may alter the pro- or antioxidant activity observed at a given moisture level. Similarly, the particular "leg" or portion of the hysteresis loop involved can greatly affect the results obtained (Labuza *et al.*, 1972). For example, at similar a_w levels, peroxide values (oxidation) increase much more rapidly if the a_w level (in this case 0.75) is attained by desorption, whereas if a similar a_w level is obtained by absorption, the onset of autoxidation is delayed markedly.

Effect of a_w on the Oxidation of Foods

The primary purpose of food preservation and storage is to maintain food products in a wholesome and acceptable condition until they are consumed. Obviously, the conditions existing during the storage period are important in determining the maximal time that a food can be held without undergoing deleterious changes.

The oxidative mechanism studies, noted in the previous section, usually dealt with model systems of one type or another. As stated earlier, studies of this type are extremely useful in establishing a basis for predicting the effects of food moisture conditions on the time of induction and rate of oxidative rancidity. An extrapolation of the model systems data to food systems predicts that, while a low a_w food might be refractory to microbial growth, enzymatic degradation, nonenzymatic browning, and other moisture-dependent deteriorative effects, extremely low moisture levels might render the food optimally susceptible to oxidative rancidity. Addition of moisture to this food should produce an antioxidative effect until a transition point is reached at which it is obscured by a prooxidant effect, which continues to increase through the IMF range and above.

Knowledge of the a_w levels at which these alterations occur would permit predictions based on the various factors affecting the storage of a particular product, the optimal a_w level for its stability. Additional benefits that might accrue from such knowledge are less expensive packaging materials, the omission of nitrogen or other inert gas packing techniques, the obviation of antioxidant additives, and the possible use of less saturated lipids which might better serve a specific food-processing purpose.

A number of foods have been examined for the influence of moisture on their development of oxidative rancidity. The most difficult foods to stabilize by water manipulation might be expected to be pure or relatively pure fats. In the case of lard, progressive increases in the water content did not alter appreciably the storage life of this product at levels below 12% (Lips, 1949). At 12% water, lard appeared to oxidize more rapidly. Chicken fat exhibited a similar acceleration in

TABLE 4.1.

Effect of Variation in Flake Moisture Content and Oven-Drying Temperature on the Rate of Development of Oxidative Rancidity in Cooked Oat Flakes[a]

Oven-drying temp. (°C)	Moisture content to which flakes were dried (%)	Stability (days)	Peroxide values at point of rancidity
180	0.9	5	129
	5.9	11	70
	9.4	150	23
140	0.1	13	130
	4.9	26	63
	9.4	134	18
110	1.0	13	245
	4.2	58	44
	8.6	250	13
Unheated, raw flakes	2.8	69	76

[a] Martin, 1958.

oxidation rate in the presence of high moisture levels, but the extent of autoxidation at or near the BET monolayer was not reported.

Many foods, such as dried cereal-based products, while not normally considered "fatty" foods, may contain appreciable levels of lipids. In such instances, the storage quality of the dried food depends greatly on its resistance to autoxidation and indirectly on the moisture condition of the cereal. For example, Martin (1958) found the stability of oat flakes to increase markedly (Table 4.1) with the moisture content of the dried flakes. In this instance, final moisture content is more influential in stabilizing the lipid fraction than the oven drying temperature. Unheated raw flakes always were more stable than low-moisture dried flakes, probably because natural antioxidants were inactivated during drying. Spray-dried milk also appears to follow the autoxidation pattern predicted by model systems studies. Loncin *et al.* (1968) noted extremely rapid increases in peroxide values of this product at an a_w of 0.00. However, if the a_w was increased to 0.18 and above (Fig. 4.3), an appreciable increase in oxidative stability occurred. Also, as predicted by model systems studies, samples poised at high a_w levels (0.75) appeared to be no more stable than much lower a_w samples.

The harvesting and subsequent storage of certain nut meats, such as shelled pecans (0.30 a_w) and, in some cases, shelled walnuts (0.17 a_w) entail special a_w-related problems with respect to oxidative rancidity. Nuts normally are steeped in water or steamed to assure more uniform breakage and better recovery of premium-priced, half-nut meats. During steeping, some moisture penetrates

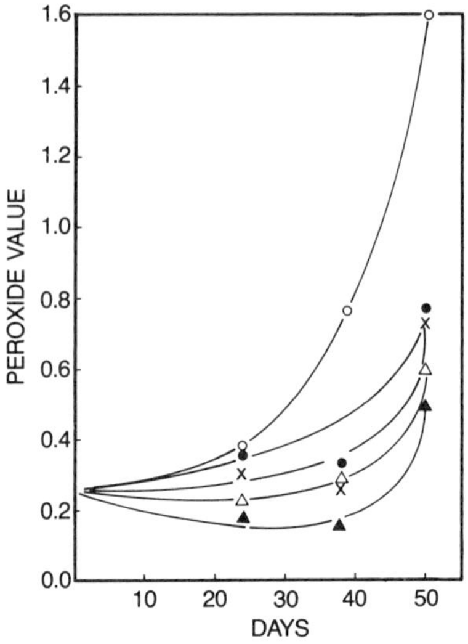

Fig. 4.3. Oxidation rate at several a_w levels of the fat contained in spray-dried milk powder at 37°C: ○, 0.00; ●, 0.415; ×, 0.75; △, 0.18; ▲, 0.53 (Loncin *et al.*, 1968).

the shell and may equilibrate the nut meats to higher a_w levels. This readjustment in some instances contributes to rancidity and off-flavor problems during extended storage. The use of hypochlorite in the steep may exacerbate this pro-oxidative condition.

Meat products may be stabilized if the water level is manipulated to minimize autoxidation. Free radical concentration directly affects the development of rancidity, and, as Munday *et al.* (1962) showed, increases in the water content of freeze-dried minced beef reduced the free radical content of this product. Interestingly, these workers attributed the stabilization of dried beef (by increasing moisture) to water molecules dislodging oxygen from the meat tissues, with the result that free radicals, normally formed by the entrapment of oxygen in tissue interstices, could not be detected. While interstitial oxygen may indeed influence the free radical level of meats, and thus the oxidative rancidity, this hypothesis seems to ignore the solubility of oxygen in water and the presence and solubility of metal-containing compounds in meats, all of which would tend to increase autoxidation. A more plausible hypothesis for the protective effect of water may be proposed from the studies of Martinez and Labuza (1968) on freeze-dried salmon. While peroxide values of this product varied inversely with relative humidity, a poor correlation between peroxide values and absorbed oxygen was found, probably as a result of the decomposition of hydroperoxides by a number of possible mechanisms. This destruction of hydroperoxides tends to inhibit bimolecular decomposition. In addition, there exists the possibility that the effect of metal catalysts present in the meat is inactivated, for example by dilution.

Salwin (1962) investigated the oxidative stability of freeze-dried pork chops and chicken and found that maximal stability occurred in both cases at the "a_l value" or water content equivalent to the calculated monomolecular water layer (9.5% and 5%, respectively, for pork and chicken). Similar results were obtained for dehydrated potato granules and nonfat dried milk. The inclusion in the latter case of in-package desiccants reduced the oxidative stability of the product.

The implications of a_w level on the autoxidation of foods preserved in the intermediate moisture food (IMF) range (a_w = 0.60–0.85) are obvious. This a_w range occurs above the transition point (a_w of least oxidation) of these foods and lies directly on that portion of the oxidation curve in which increasing amounts of moisture rapidly increase oxidation rates. It is, thus, imperative that foods equilibrated to moisture levels in the IMF range be formulated with maximal stability to oxidative rancidity as a consideration. One method of solving this problem is to utilize antioxidants. Another, not necessarily exclusive of the first, is to exploit the differences in the adsorption and desorption curves of the hysteresis loop. Labuza *et al.* (1972) have shown that foods prepared by adsorption techniques oxidize much more slowly than those formulated via desorption. Although the former requires that a food be dried below the IMF range and then equilibrated upward, a costly process, the advantages accruing from the postponement of oxidative rancidity may justify the additional expense.

Antioxidants

The principal role of antioxidants in foods subject to oxidative rancidity is one of quenching free radical propagation [Eq. (3)] and so interrupting the chain of oxidative reactions. Thus, these compounds (XH, Eq. 4) act as free radical scavengers, as indicated in Eq. 4.

$$\underset{\text{}}{R^\cdot} + \underset{\substack{\text{anti-}\\\text{oxidant}}}{XH} \longrightarrow RH + \underset{\substack{\text{stable}\\\text{radical}}}{X^\cdot} \qquad (4)$$

As noted earlier in this chapter, many foods contain natural antioxidants such as α-tocopherol. However, these materials are often not very effective and are costly (if added to foods), and in many cases, are readily inactivated by even mild heat treatments. As a result, more effective phenolic antioxidants have been synthesized and are widely utilized in the food industry. Butylated hydroxyanisole (BHA), butylated hydroxytoluene (BHT), and propyl gallate (PG) are examples.

In practice, metal-chelating agents, such as citric or ascorbic acid, may be used in conjunction with the phenolic antioxidants. In this situation, a synergistic effect is obtained, attributed by Privett (1961) to a sparing action on the antioxidant.

The effectiveness of antioxidants varies with the amount of water present in the system, a factor which must be considered in assessing their applicability to foods subject to autoxidation. Labuza *et al.* (1969) found in methyl linoleate-containing model systems that the antioxidant activity of PG increases at higher a_w, whereas BHT is more effective in "dry" systems.

In many instances, the water solubility may determine, to a large degree, the effectiveness of antioxidants such as ethylenediaminetetraacetic acid (EDTA) or citric acid. Increases in water activity enhance the antioxidant effectiveness of both of these compounds. EDTA tends to bind to proteins, and thus its effectiveness in many foodstuffs is reduced. The phenolic antioxidants BHT and BHA do not appear to be functionally impaired in similar circumstances.

In addition, just as the method of obtaining a given moisture level (hysteresis) influences the effect of a_w on autoxidation, so does it alter the effectiveness of antioxidants. Chou and Labuza (1974) found that antioxidants tend to be more effective in adsorption-prepared systems than in desorption systems. These authors related the antioxidant "effectiveness ratio" (ratio of induction time in the presence of antioxidant to induction time in a control) to the a_w of the system, type of food, and condition of preparation (absorption or desorption). A portion of these data is found in Table 4.2 and shows that an increase in a_w from 0.75 to 0.84 usually produces an increase in the effectiveness of a variety of antioxi-

TABLE 4.2.
Antioxidant Effectiveness in a Methyl Linoleate–Cellulose Model System Supplemented with 1000 ppm Metals[a]

		Effectiveness ratio[b]	
Antioxidant	a_w	Desorption	Adsorption
EDTA	0.75	22.5	10.7
(0.08 moles/mole metal)	0.84	12.4	10.8
Sodium citrate	0.75	2.8	2.4
(0.5 moles/mole metal)	0.84	2.2	2.2
Isopropyl citrate	0.75	2.6	2.1
(0.5 moles/mole metal)	0.84	1.6	2.0
α-Tocopherol	0.75	1.6	1.4
(200 ppm, fat basis)	0.84	1.3	1.2
BHA (200 ppm, fat basis)	0.84	4.6	5.6

[a] Chou and Labuza, 1974.
[b] Ratio of induction time to induction time of respective control.

dants. These results generally characterize the effectiveness of antioxidants in systems of various a_w levels.

EFFECT OF a_w ON FOOD TEXTURE

In some respects, the influence of moisture content on the perception of food texture is obvious. Usually this is based on the "toughness" of a food and is related to the type of food consumed. For example, cooked beef may be relatively tough, yet have a higher a_w level than honey or syrups, which are perceived as viscous liquids. On the other hand, various samples of beef muscle tissue may differ considerably in texture, depending on numerous factors such as the protein-to-fat ratio, type of meat protein involved, temperature of cooking, and moisture content, both before and after cooking. In this more subtle relationship, the effects of moisture may be evaluated in terms of protein or carbohydrate hydration.

Evaluation of the effect of relative humidity or a_w on texture has been plagued by the difficulties of objectively quantifying texture. Although various tests have been applied, this characteristic remains essentially a subjective attribute. These restrictions notwithstanding, a number of studies have shown convincingly that the moisture condition of dried or semidried foods plays an important role in determining texture. Kapsalis (1967) found that increases in equilibrium relative humidity of intermediate moisture foods above the BET monolayer of water (20% relative humidity) to a moisture level of 66% produced increases in both hardness and cohesiveness of most samples. In the case of the meat samples examined, this increase in toughness was related to cross-linking and possibly to other chemical reactions known to occur at humidities greater than the monolayer level. Colloidal aggregation and changes in hydration among different sorptive groups were also mentioned as possible mechanisms. In a more recent study, Heldman *et al.* (1972) found that Instron measurements (an instrumental method for determining hardness and chewiness) of precooked, freeze-dried beef increase as the relative humidity is increased, until a moisture content in the 40–50% R.H. range is reached (Fig. 4.4). Further increases in moisture decreased both hardness and chewiness factors. Results for these test parameters were similar at 80% and 0% R.H.

Chordash and Potter (1972) investigated the feasibility of controlling texture in a number of foods by varying moisture. In all of the foods (ham, beef, custard, and peas), an increase in a_w from 0.68 or 0.70 to > 0.99 resulted in an increase in desirable subjective texture attributes. These results were somewhat compromised by bacterial growth at the higher a_w levels tested (> 0.99 and 0.98), which could have altered the textural parameters.

At present, the most meaningful data relating moisture to texture have been

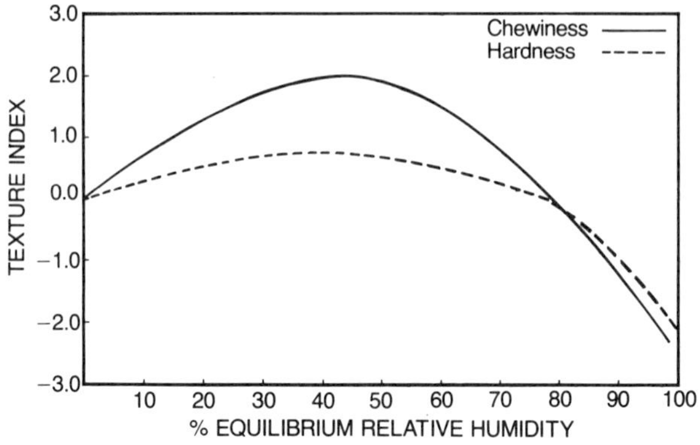

Fig. 4.4. Statistical-fit plots of effect of relative humidity on texture (Instron readings) of freeze-dried beef (plotted from data of Heldman *et al.*, 1972).

derived from studies with dried and semidried meat. Clearly, the study of this relationship is in its infancy. The need to establish objective data has, perhaps, been emphasized in this area to the detriment of the subject itself. In an attempt to provide usable methodology, instrument manufacturers have developed machines, some almost whimsical in design and function, which provide data on such parameters as toughness, crispness, modulus of elasticity, bioyield strength, degree of elasticity, compression, penetration, and a host of other seemingly confusing terms. Perhaps the most meaningful data may evolve from the human mouth itself via the use of subjective panel tests to obtain hedonic texture grades. In the final analysis, what one perceives as the texture of a food is a specific experience unique to the beholder.

EFFECT OF a_w ON FOOD PIGMENTS

The principal effect of moisture condition on food color development was discussed in the section of the previous chapter that deals with browning reactions. A very few reports also exist on the effect of a_w on natural plant and animal pigments, principally astacene (the pink pigment found in salmon) and chlorophyll.

Chlorophyll degradation in a model system and in blanched, freeze-dried spinach was studied by LaJollo *et al.* (1971), who found that a_w influences chlorophyll decomposition in both cases. In spinach, degradation of chlorophyll to the gray pigment, pheophytin, was relatively slow at 0.32 a_w and below, but increased at levels greater than 0.32. Assuming that the product would be acceptable when 20% or less of the chlorophyll is degraded to pheophytin (the concentration at which the gray color begins to predominate), these authors then related a_w to time required for this degradation. Twenty days were required to produce the 20% loss at 0.32 a_w, whereas at 0.75 a_w it occurred in less than 2 days. In the model system, chlorophyll conversion to pheophytin proceeded very rapidly, with the reaction rate increasing under oxidizing conditions. Although a_w probably is important in the stabilization of chlorophyll, other factors such as pH, autoxidation, and nonenzymatic browning appear to interact. The color stability of green peppers has been reported (Salwin, 1962) to be greatest at the monolayer water content (1.6% for this vegetable). Water content above or below this level led to a more rapid breakdown of the pigment. Another chlorophyll-containing vegetable, cabbage, was stored for 6 months at 1°C at 100, 85, and 75% relative humidity levels by Pendergrass and Isenberg (1974). Assuming that ample time was allowed for equilibration of the cabbage to the stated relative humidities, color losses were much less at 100% relative humidity, these authors recommended high humidity storage for this product.

The adjustment of freeze-dried, precooked carrots to various a_w levels, on the other hand, resulted in complete destruction of carotene at 0.08 a_w, and increasing stability of this pigment was observed up to 0.30, the highest a_w studied. In both carrots and sweet potatoes, drying below the BET monolayer level resulted in oxidation of the carotene to betaionone, a violet pigment, within 37 days. It should be pointed out, however, that carotene is readily decomposed by lipid hydroperoxides and that, in such instances, the a_w of the system may be of only incidental importance.

Recent U.S. regulatory restrictions on the use of artificial dyes in foods have stimulated interest in natural plant pigments as alternative food colorants (Pasch and von Elbe, 1975). As a result of this interest, work has been pursued relating to those factors which effectively stabilize the red pigments in beets, most notably betanine. The half-life of this pigment has been found to be greatly increased at 0.47 a_w as compared to 1.00. Exploitation of a_w control and other factors has been suggested as means of extending the stability and usefulness of beet-derived pigments.

The pigments of freeze-dried salmon were studied by Martinez and Labuza (1968), who found that moisture conditions above the monolayer gave almost total stability to the pigments involved, primarily astacene. In this case, as with carotene (as noted above), oxidative stability appeared to be involved, probably resulting from the inactivation of metal ions by hydration.

Although data on food pigment stability as related to a_w are few, chlorophyll probably behaves differently from other pigments of plant or animal origin. While the latter pigments appear to be stabilized with increasing a_w, chlorophyll, paradoxically, becomes less stable. Theories exist on the basis of this difference, but thus far definitive work has not been published.

EFFECT OF a_w ON NUTRIENTS

Vitamins

Most of the work done on the influence of a_w on the nutritional composition of foods has been concerned with ascorbic acid. This vitamin is relatively stable at low a_w levels, but increases in the water content of model systems and foods result in rapidly increasing rates of ascorbic acid decomposition (Fig. 4.5). It appears that this vitamin is more rapidly destroyed in desorption than adsorption systems (Labuza, 1973; Lee and Labuza, 1975). In addition to a_w, the temperature of the reaction also strongly influences the rate of destruction. Nuclear magnetic resonance studies on the aqueous phase of model systems indicate that the increased rate of destruction may be related to decreasing viscosity. Similar findings have been reported with dehydrated orange juice by Karel and Nickerson (1964). They found that little decomposition occurred at 0.01 a_w, but that rates increased in samples equilibrated to higher a_w levels until, at 0.53 a_w, ascorbic acid could not be detected after approximately 8 weeks of storage. These data may be unique because the curves relating decomposition to a_w are linear throughout the test range and extrapolate to the origin. This indicates that ascorbic acid decomposition may proceed at and below the BET monolayer.

The preservation of ascorbic acid in foods for extended periods, therefore, requires that the food be stored at low temperatures and at as low a_w levels as are

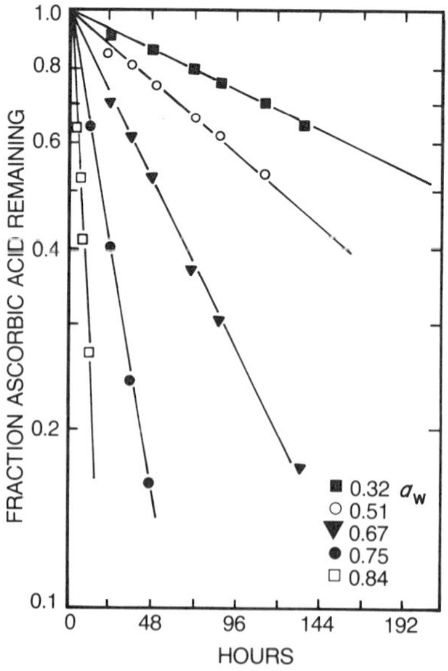

Fig. 4.5. Ascorbic acid destruction as a function of a_w at 35°C (Labuza, 1974).

practicable. Other methods which have been suggested by Labuza (1974) to stabilize ascorbic acid in vitamin-supplemented foods are to protect it from moisture by encapsulation or by adding it to the oil phase of the product.

Requirements in many countries for the vitamin supplementation of wheat flour have attracted much attention to the stability of thiamine in such products. Extra vitamin must often be added to compensate for losses occurring during extended storage under a variety of conditions. As a result, much has been learned about the loss of thiamine in stored cereal products. In one of the earliest reports of these tests, Hollenbeck and Obermeyer (1952) demonstrated that the type of thiamine salt strongly influences the extent of vitamin loss.

In these studies, thiamine chloride hydrochloride and thiamine mononitrate were added to flour equilibrated to various moisture levels. After 4 months' storage, vitamin loss was greater at 38°C than at 28°C. Increases in flour moisture from 9.2 to 14.5% resulted in increased loss of both forms of thiamine at 38°C, especially the hydrochloride. Other evidence that thiamine losses are accelerated by moisture increases has been provided by Cuendet *et al.* (1954), who stored several wheat products containing 3–14% water at 38°C. Consistent losses of up to 80% of the thiamine content occurred in the 14% samples, whereas the thiamine content of flours with moisture contents of 3, 6, and 10% remained unchanged.

Data pertaining to thiamine loss in meats as a function of water content have been restricted largely to studies on dehydrated pork (Rice *et al.*, 1944), which showed that degradation of this vitamin is proportional to water content in the 0–6% range (Table 4.3). Canned pork (55% water) was more stable than dehydrated pork (2% water); therefore, other factors must be involved. It is also interesting to note that the addition of 5% NaCl did not stabilize the thiamine content appreciably. However, it is difficult to ascertain the a_w levels of these systems.

The mechanism by which thiamine is stabilized at low moisture levels is not understood. Labuza (1973) has suggested that browning reactions might be in-

TABLE 4.3.

Influence of Moisture Content of Dehydrated Pork on Thiamine Stability[a]

Moisture (%)	Thiamine retained 7 days at 49°C
0	91
2	60
4	23
6	9
9	11

[a] Rice *et al.*, 1944.

volved. Kinetic studies indicate that the degradation of thiamine follows first-order reaction kinetics, exhibiting a very high activation energy in aqueous solutions.

The destruction of fat-soluble vitamins appears to be related to lipid oxidation and generally decreases with water activity. The exception to this generalization may be α-tocopherol (vitamin E), which has been reported to undergo accelerated decomposition with increased moisture content (Jensen, 1969).

Protein Quality

Browning or Maillard reactions probably account for much of the loss in the nutritional quality of protein that may occur as the moisture content of foods increases. On the other hand, if lipids susceptible to autoxidation are present in a proteinaceous food, free radical–protein reactions may occur which could cause losses. In this case, one would expect increasing a_w to have a stabilizing effect. Probably protein oxidation, which occurs at very low a_w levels, could be stabilized by antioxidants (Roubal, 1970).

The above reports and theories pertain primarily to the overall effects of a_w on protein quality as measured by protein efficiency ratio. The effect of a_w on the fate of individual amino acids in foods is virtually unknown, with the somewhat dubious exception of lysine loss in browning reactions, discussed in the previous chapter. Clearly, there is a need to examine in greater detail the effects of a_w on the nutritional quality of foods. Data thus far reported suggest that the subject is complex.

REFERENCES

Chordash, R. A., and Potter, N. N. (1972). Effects of dehydration through the intermediate moisture range on water activity, microbial growth, and texture of selected foods. *J. Milk Food Technol.* **35,** 395–398.

Chou, H. E., and Labuza, T. P. (1974). Antioxidant effectiveness in intermediate moisture content model systems. *J. Food Sci.* **39,** 479–483.

Cuendet, L. S., Larson, E., Norris, C. G., and Geddes, W. F. (1954). The influence of moisture content and other factors on the stability of wheat flours at 37.8°C. *Cereal Chem.* **31,** 362–389.

Heidelbaugh, N. D., and Karel, M. (1970). Effect of water-binding agents on the catalyzed oxidation of methyl linoleate. *J. Am. Oil Chem. Soc.* **47,** 539–544.

Heidelbaugh, N. D., Yeh, C. P., and Karel, M. (1971). Effects of model system composition on autoxidation of methyl linoleate. *J. Agric. Food Chem.* **19,** 140–142.

Heldman, D. R., Bakker-Arkema, F. W., Naoddy, P. O., Reidy, G. A., Paltnicker, M. P., and Thompson, D. R. (1972). Investigation of the energetics of water binding in dehydrated foods at very low moisture levels in relation to quality parameters. *U.S., Army Natick Lab. Tech. Rep. 72-10-FL.*

Hollenbeck, C. M., and Obermeyer, H. G. (1952). Relative stability of thiamine mononitrate and thiamine chloride hydrochloride in enriched flour. *Cereal Chem.* **29,** 82–87.

Jensen, A. (1969). Tocopherol content of seaweed and seaweed meal. III. Influence of processing and storage on the content of tocopherols, carotenoids, and ascorbic acid in seaweed meal. *J. Sci. Food Agric.* **20,** 622–626.

Kapsalis, J. G. (1967). Hygroscopic equilibrium and texture of freeze-dried foods. *U.S., Army Natick Lab. Tech. Rep. 67-87-FL.*

Karel, M., and Nickerson, J. T. R. (1964). Effects of relative humidity, air, and vacuum on browning of dehydrated orange juice. *Food Technol. (Chicago)* **18,** 1214–1218.

Labuza, T. P. (1971). Kinetics of lipid oxidation in foods. *Crit. Rev. Food Technol.* **2,** 355–402.

Labuza, T. P. (1973). Effects of dehydration and storage. *Food Technol. (Chicago)* **27,** 20–26 and 51.

Labuza, T. P. (1974). Storage Stability and Improvement of Intermediate Moisture Foods, Final Rep., NAS Contract 9-125-60, Phase II, pp. 10–81. Natl. Acad. Sci., Washington, D.C.

Labuza, T. P., Maloney, J. F., and Karel, M. (1966). Autoxidation of methyl linoleate in freeze-dried model systems. II. Effect of water on cobalt-catalyzed oxidation. *J. Food Sci.* **31,** 885–891.

Labuza, T. P., Tsuyuki, H., and Karel, M. (1969). Kinetics of linoleate oxidation in model systems. *J. Am. Oil. Chem. Soc.* **46,** 409–416.

Labuza, T. P., Heidelbaugh, N. D., Silber, M., and Karel, M. (1971). Oxidation at intermediate moisture contents. *J. Am. Oil Chem. Soc.* **48,** 86–90.

Labuza, T. P., McNally, L., Gallagher, D., Hawkes, J., and Hurtado, F. (1972). Stability of intermediate moisture foods. 1. Lipid oxidation. *J. Food Sci.* **37,** 154–159.

LaJollo, F., Tannenbaum, S. R., and Labuza, T. P. (1971). Reaction at limited water concentration. 2. Chlorophyll degradation. *J. Food Sci.* **36,** 850–853.

Lee, S. H., and Labuza, T. P. (1975). Destruction of ascorbic acid as a function of water activity. *J. Food Sci.* **40,** 370–373.

Lips, H. J. (1949). Effect of added water and antioxidants on the keeping quality of lard. *Can. J. Res.* **27,** 373–381.

Loncin, M., Bimbenet, J. J., and Lenges, J. (1968). Influence of the activity of water on the spoilage of foodstuffs. *J. Food Technol.* **3,** 131–142.

Maloney, J. F., Labuza, T. P., Wallace, D. H., and Karel, M. (1966). Autoxidation of methyl linoleate in freeze-dried model systems. I. Effect of water on the autocatalyzed oxidation. *J. Food Sci.* **31,** 878–884.

Martin, H. F. (1958). Factors in the development of oxidative rancidity in ready-to-eat crisp oatflakes. *J. Sci. Food Agric.* **9,** 817–824.

Martinez, F., and Labuza, T. P. (1968). Rate of deterioration of freeze-dried salmon as a function of relative humidity. *J. Food Sci.* **33,** 241–247.

Munday, K. A., Edwards, M. L., and Kerkut, G. A. (1962). Free radicals in lyophilised food materials. *J. Sci. Food Agric.* **13,** 455–458.

Pasch, J. H., and von Elbe, J. H. (1975). Betanine degradation as influenced by water activity. *J. Food Sci.* **40,** 1145–1146.

Pendergrass, A., and Isenberg, A. F. M. (1974). The effect of relative humidity on the quality of stored cabbage. *Hortic. Sci.* **9,** 226–227.

Privett, O. S. (1961). Some observations on the course and mechanism of autoxidation and antioxidant action. *Proc. Flavor Chem. Symp., 1961* pp. 147–161.

Rice, E. E., Beuk, J. F., Kauffman, F. L., Schultz, H. W., and Robinson, H. E. (1944). Preliminary studies on stabilization of thiamine in dehydrated foods. *Food Res.* **9,** 491–499.

Roubal, W. T. (1970). Trapped radicals in dry lipid–protein systems undergoing oxidation. *J. Am. Oil Chem. Soc.* **47,** 141–144.

Salwin, H. (1959). Defining minimum moisture contents of dehydrated foods. *Food Technol. (Chicago)* **13,** 594–595.

Salwin, H. (1962). Moisture in deteriorative reactions of dehydrated foods. *Freeze-Drying Foods, Proc. Conf., 1961* pp. 58–74.

5

Microbial Growth

The changes that microorganisms may effect in the quality of foods are considered as contributions to food preservation or to food spoilage, depending upon whether the changes are organoleptically desirable or undesirable. Whatever the nature of the changes, it is clear that they only reach readily observable proportions when the responsible microorganisms are present in the food in large numbers. Such large numbers are sometimes introduced as inocula in food fermentations. However, most commonly the initial population or microbial load is small and reaches levels that produce these changes only after extensive multiplication in the food.

The development of microbial cultures takes slightly different forms in bacteria, yeasts, and molds, and varies also as a consequence of germination and sporulation, where these occur. Whether growth ensues from binary fission, budding, or by hyphal extension, the phases through which the cultures pass are broadly similar. A period of adjustment or adaptation (the lag phase) is followed by accelerating growth until a steady, rapid rate is achieved. Later growth slows until growth and death are balanced and the population remains constant (the stationary phase). Eventually, death exceeds growth and the culture enters the phase of decline.

In bacteria and unicellular fungi, these phases are usually well defined. Growth during the rapid phase is exponential, and a logarithmic plot of the population as a function of time yields a straight-line relationship. From this, the generation time or, conversely, the number of doublings occurring in a unit period (the growth rate), may be determined.

It is customary to record the growth of filamentous fungi on solid media in terms of the increase in colony radius or diameter. These measurements give a growth curve that is initially and briefly exponential and, subsequently, arithmetic. In submerged liquid culture, however, the kinetics of growth, based on dry weight measurements, are frequently similar to those of unicellular microorganisms in being predominantly exponential.

Water activity can be shown to influence each of the four main growth cycle phases by its effects upon the germination time or the length of the lag phase, the growth rate, the size of the stationary population, and the subsequent death rate. Experimentally, the parameters most often recorded are the rate of growth, the extent of growth (the size of the stationary phase population), and the presence or absence of growth, which is a summation of effects on the lag, growth rate, and population reached. Quantitative examination of the phase of decline is less commonly performed on the growth culture than on microorganisms transferred to other environments.

Clearly, the most important criterion in the water relations of a particular microorganism is the minimal a_w permitting growth. However, from the point of view of food technology and preservation, less extreme effects also may be of value. For instance, at some a_w levels above this minimum, the population of a pathogen may be insufficient to produce an infectious dose or a toxic product. More importantly, any reduction in the rate of growth resulting from reduced a_w may pave the way to synergistic inhibition by another physical or chemical agent, e.g., pH or curing salts. It is appropriate, therefore, to consider the effects of a_w on several aspects of microbial growth and survival.

BACTERIA

Bacteria are generally the fastest growing of microorganisms, so that in conditions favorable to both, they will usually outgrow fungi. They are important, as agents of both spoilage and food-borne disease, and the extensive studies which have been performed on the water relations of the latter, pathogenic group are discussed in Chapter 8. Here we shall consider the influence of water activity on bacteria in general.

Most nonhalophilic bacteria have maximal growth rates at a_w levels in the 0.997–0.980 range. The most dilute laboratory media in common use, e.g., nutrient broth, are about 0.999 a_w, and adding solutes to reduce a_w to about 0.995 will increase the growth rate of members of the *Enterobacteriaceae* by 10–15%.

At a_w levels below the optimum, growth rate declines, often linearly. The decline will be large for organisms of high water requirement and small for those capable of growth under conditions of physiological drought. In contrast, the stationary population remains at a high, often constant level until an a_w that reduces the growth rate by as much as 50% is reached.

The a_w levels giving maximal growth rates are often assumed to support the shortest lag phases, although it is usually difficult to measure lag accurately at high a_w. At values close to the minimum for growth, the lag is, of course, extended. The viable population may decrease by 90% or more before multiplication commences, and a long period may elapse before it returns to the inoculation level.

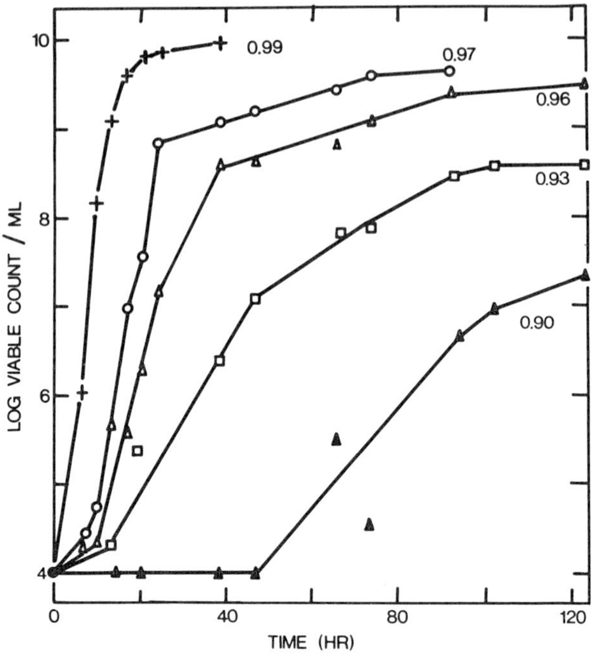

Fig. 5.1. Growth curves of *Staphylococcus aureus* C-243 in a laboratory medium adjusted to various a_w levels with hydrolyzed protein (Troller, 1971).

The water relations of the lag phase differ greatly among species. In some cases, the lag begins to lengthen immediately after the a_w is reduced below the optimum for multiplication. In a different species of the same genus, conditions that halve the growth rate may have little effect on the duration of the lag phase. Examples of bacterial growth curves determined by Troller (1971) over a range of a_w values are shown in Fig. 5.1.

There are few data to indicate how the yield of cells or the stationary population is influenced by very high a_w levels, but clearly, if these levels are obtained by dilution of nutrients, there will be a threshold above which the yield will decline.

Many more data are available on the minimal water activities permitting growth of bacteria than on the kinetic aspects of bacterial growth referred to above. A selective summary of published information on limiting a_w levels for various microorganisms is given in Appendix B.

In general terms, among nonhalophilic bacteria, gram-negative rods are most sensitive to reduced a_w, having minima for growth in laboratory media in the range 0.96–0.94. For *Clostridium* spp., the minimum is frequently 0.95–0.94

and for *Bacillus* spp., 0.93–0.90 a_w. The widest spread is that shown by gram-positive cocci, which have been reported to have a_w limits ranging from a minimal a_w of 0.95–0.83.

Water relations quite different from those described above are found among halophilic bacteria. In addition to being generally tolerant of high concentrations of salt and hence, growing in environments of low a_w, they are unable to grow at high a_w levels. The low a_w necessary for growth must be provided by inorganic salts, which must in turn be predominantly NaCl.

The salt relations of relatively few such isolates have been studied. These fall into two groups, the moderately and extremely halophilic, but it may be that a continuum of salt relations exists in nature.

The most studied moderate halophiles require 0.2–0.5 *M* NaCl for growth, grow optimally in about 1 *M* NaCl, and will grow at concentrations up to 3.5–4.0 *M* NaCl. These concentrations are roughly equivalent to 0.99, 0.96, and 0.85 a_w. Extreme halophiles grow best at 4.0 *M* NaCl (0.83 a_w), with a growth range of 3 *M* (0.88 a_w) to saturated NaCl (0.75 a_w). For moderately halophilic bacteria, some 80–90% of the apparent NaCl requirement can be replaced by KCl, while for extreme strains, the replaceable concentration is 60–70%.

It follows that although these halophiles are adapted to grow at very low levels of a_w, they are very sensitive to the solute responsible for reducing a_w. Consequently, they are of concern only in that group of foods preserved predominantly with NaCl. The influence of solute upon the water relations of less exotic organisms will be referred to later.

FUNGI

In nature, yeasts and molds are commonly found to grow either in more acid conditions or in drier or more concentrated environments than bacteria. Hence, the a_w levels required to prevent growth of most fungi are lower than those preventing multiplication of most bacteria. The very drought-resistant fungi have been termed xerophilic molds and osmophilic yeasts, although the distinction is probably illusory. As an arbitrary definition, these groups will be taken to include all species of fungi "capable of growth, under at least one set of environmental conditions, at a water activity below 0.85" (Pitt, 1975).

Yeasts

Species of yeasts which are usually found growing where the a_w is high, such as those used to produce alcohol, nevertheless can multiply over a wide range, commonly to levels near 0.90 a_w. The influence of a_w on the growth rate of one such yeast, *Saccharomyces cerevisiae*, is shown in Fig. 5.2. The trend of increas-

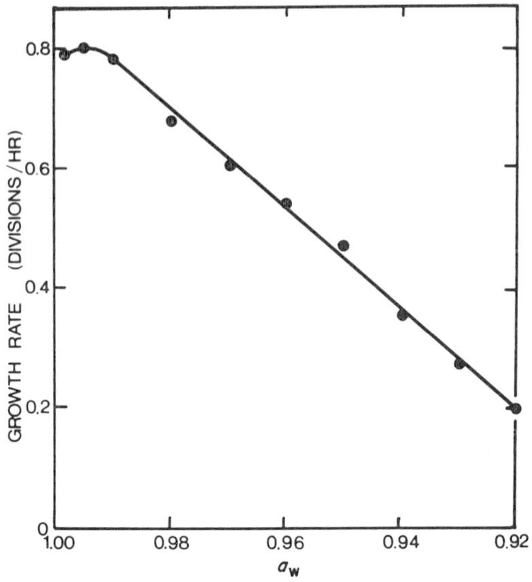

Fig. 5.2. Effect of a_w on growth rate of *Saccharomyces cerevisiae* using sucrose to control a_w. (J. H. B. Christian, unpublished data.)

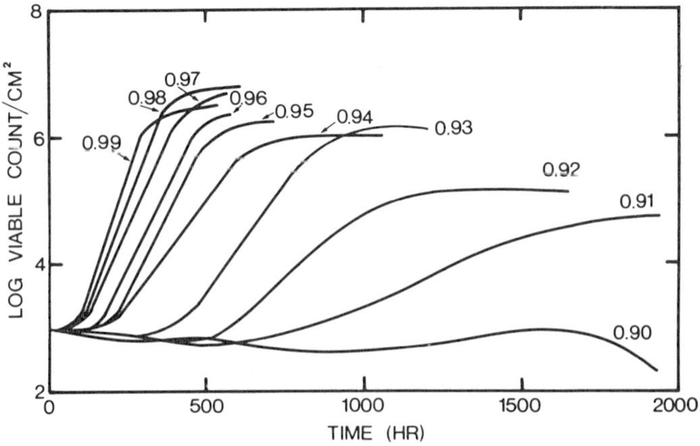

Fig. 5.3. Growth curves of *Geotrichoides* sp. Y9 at −1°C on ox muscle slices adjusted to various a_w levels by equilibration (Scott, 1936a).

ing lag, decreasing multiplication rate, and decreasing stationary population is depicted clearly for a *Geotrichoides* sp. in Fig. 5.3 (Scott, 1936a).

Most of the data existing on the water relations of yeasts refer to osmophilic (or xerophilic) species. Most osmophiles appear to be isolates of *Saccharomyces rouxii*, reported by von Schelhorn (1950) to grow to a minimal a_w of 0.62 in a fructose syrup and a pear concentrate. Strains of *S. rouxii* are also among the most salt-tolerant yeasts, playing an important role in the ripening of soy mashes. Growth may occur at levels close to 0.81 a_w. *Saccharomyces, Pichia, Debaryomyces, Torulopsis,* and *Hansenula* are among the genera most commonly reported as containing highly salt-tolerant species (Ōnishi, 1963). No yeast appears to have been reported as growing in substrates saturated with NaCl (0.75 a_w).

Molds

Many of the molds, such as some species of *Mucor, Neurospora,* and *Rhizopus,* which grow rapidly on moist foods, are of no concern when the a_w level is much below 0.90 a_w, and spores of some plant pathogenic species will

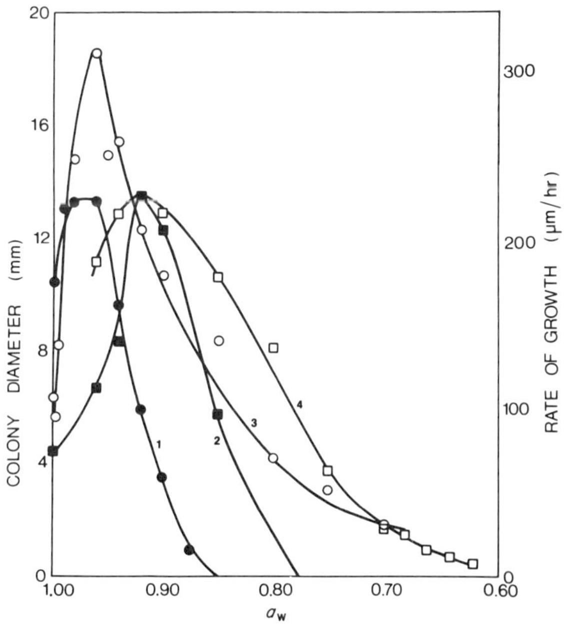

Fig. 5.4. Relationship between rates of growth and a_w for four molds. Curve 1: left ordinate *Aspergillus niger* at 20°C. Curve 2: left ordinate *A. glaucus* at 20°C. Curve 3: right ordinate *A. amstelodami* at 25°C. Curve 4: right ordinate *Xeromyces bisporus* at 25°C (Scott, 1957).

germinate only above 0.99 a_w. As with yeasts, the concern here is largely with the xerophilic species, those growing below 0.85 a_w.

Xerophilic molds have representatives in many genera, but predominantly among *Aspergillus* (including *Eurotium*) and *Penicillium* spp. Pitt (1975) reviewed the water relations of some 40 xerophilic molds, 7 of which were reported to grow at a_w levels of 0.70 or below: *Aspergillus conicus* (0.70 a_w), *Chrysosporium fastidium* (0.69 a_w), *Eremascus albus* (0.70 a_w), *Eurotium amstelodami* (0.70 a_w), *Eurotium echinulatum* (0.62 a_w), *E. rubrum* (0.70 a_w), and *Xeromyces bisporus* (0.61 a_w). The latter is the lowest level of a_w at which microbial growth has been recorded.

Figure 5.4 illustrates the influence of a_w upon the rates of growth of four species of mold (Scott, 1957). Note that by the definition given earlier, three are considered xerophilic. All four molds show clear optimal a_w levels, which for two species are at the low level of 0.92.

As Pitt (1975) points out, all of the xerophilic molds so far reported to be mycotoxic belong to *Aspergillus* (including *Eurotium* and *Emericella*) and *Penicillium*. Although some data have been reported on environmental influences on toxin production, there are few indications of how toxigenesis is affected by reduced a_w. Of the 16 species of xerophilic molds listed by Pitt (1975) as known to produce mycotoxins, only two are capable of growth below 0.75 a_w and none below 0.70 a_w.

INTERACTIONS WITH WATER ACTIVITY

Water activity is only one of the environmental factors with which microorganisms must contend. Their growth and survival will be influenced concomitantly by temperature, hydrogen ion concentration, oxygen and carbon dioxide concentrations, and the presence of inhibitory substances, such as preservatives. Most published data on microbial water relations were obtained when these other factors were close to optimal for the organism concerned, and when chemical inhibitors were absent. When any of these other factors is suboptimal, the inhibitory effect of reduced a_w tends to be enhanced. This makes feasible the use of combinations of relatively mild conditions, which in sum will yield shelf-stable products. A typical example is the category of intermediate moisture foods, a group which is discussed in a later chapter.

Temperature

Generally, microorganisms show greatest tolerance of reduced a_w at temperatures close to their optimum for growth. A clear example, involving *Clostridium botulinum* type B, is given in Table 5.1 (Ohye and Christian, 1967). The

TABLE 5.1.

Growth of *Clostridium botulinum* Type B at Various Levels of Temperature, pH, and a_w[a]

		a_w						
Temperature	pH	0.997	0.99	0.98	0.97	0.96	0.95	0.94
20°C	5							
	6	4[b]	9	9				
	7	2	2	4	9			
	8	2	2	4	14			
	9							
30°C	5							
	6	2	2	3	9			
	7	1	1	2	3	9	14	
	8	1	1	2	4	14		
	9							
40°C	5							
	6	1	2	2	3	14		
	7	1	1	1	2	3	9	17
	8	1	1	1	2	9	14	
	9							

[a] Ohye and Christian, 1967.
[b] Incubation period (in days) before growth was observed. No growth occurred at 10°C.

optimum temperature for growth of this organism at high a_w is 37°–40°C, and the table shows that as the incubation temperature was reduced to 30°C and then to 20°C, the minimal a_w permitting growth increased markedly.

Wodzinski and Frazier (1960, 1961a–d) obtained comparable results for *Pseudomonas fluorescens* over the temperature range 15°–30°C, and for *Aerobacter aerogenes* and *Lactobacillus viridescens* at 15°–37°C. However, the picture is less clear for extremely halophilic bacteria, where increasing incubation temperature is accompanied by a substantial increase in the optimum salt concentration for growth—i.e., a reduction in the optimum a_w level. At the same time, the organism becomes less exacting in its salt requirement.

For an osmophilic yeast, von Schelhorn (1950) recorded minimal a_w values for fermentation of 0.66 at 30°C, 0.70 at 20°C and 0.78 at 10°C. Ōnishi (1963) found that *Saccharomyces rouxii* could multiply at 40°C in media containing high levels of salt or sugar, but not in their absence.

Studies by Ayerst (1966, 1969) confirm and extend earlier data that molds are most tolerant of low a_w at temperatures near the optimum for growth. For example, for a strain of *Aspergillus ruber*, growth occurred down to 0.85 a_w at 5°C, 0.80 a_w at 10°C, 0.725 a_w at 20° and 30°C, 0.75 a_w at 35°C, and 0.80 a_w at 37°C.

pH

Like temperature, this factor is a very important selective agent, and again one that influences markedly the minimal a_w at which a microorganism can grow. Table 5.1 shows that at all temperatures tested, the lowest minimal a_w for growth of *Clostridium botulinum* type B was recorded at pH 7, followed by pH 8, and then by pH 6. For fungi, the optimum pH is generally closer to 5.0, but the overall picture is similar. *Saccharomyces cerevisiae* in 0 or 1.0 *M* NaCl grew at all pH values (3.0–7.0) tested, but, in 3 *M* NaCl (0.88 a_w), only at pH 4.0 and 5.0 (Ōnishi, 1963).

Oxygen and Carbon Dioxide

Growth of microorganisms with a substantial oxygen requirement will be inhibited when the partial pressure of oxygen is low. Under such conditions, reduced levels of a_w would be expected to produce additional inhibition. Since molds require oxygen, albeit at quite low levels in some instances, this effect could be of practical importance. However, it does not seem to have been studied in the absence of such contributing factors as elevated carbon dioxide concentrations.

Bacteria have been more widely studied in this respect. Scott (1957) summarized publications showing that, while anaerobic conditions do not increase the minimal a_w for growth of salmonellae, for *Staphylococcus aureus* the anaerobic minimum is 0.90 a_w, compared with 0.86 a_w in air.

The most extensive such data for bacteria are those of Wodzinski and Frazier (1960, 1961a–d). In two-thirds of the temperature–a_w combinations tested, *Pseudomonas fluorescens*, an obligate aerobe, grew significantly more rapidly in air than in an atmosphere of 99.9% nitrogen. Although there was little difference between growth rates in air and in 10% oxygen + 90% carbon dioxide at a_w levels from 0.999 to 0.97 at 20°, 25°, and 30°C, at all temperatures growth occurred at 0.965 a_w in 10% oxygen, but not in air. At 25°C, growth was recorded at 0.965 a_w in 99.9% nitrogen, but not in air. It follows that the influence of oxygen on microbial water relations is yet to be clearly defined.

These authors showed that carbon dioxide at 5% and 10% in air significantly increased generation times at low a_w but decreased them (in 5% carbon dioxide) at higher levels. The inhibitory influence of carbon dioxide upon growth rate at various water activities is obvious at low temperatures, as shown in Fig. 5.5 for several bacteria and yeasts at −1°C (Scott, 1936b).

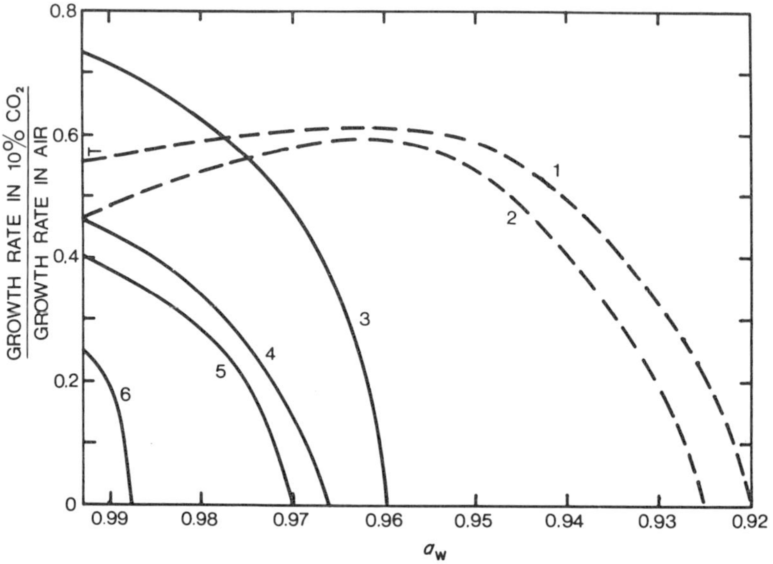

Fig. 5.5. Relation between tolerance to 10% CO_2 and the a_w at − 1°C for: (1) *Geotrichoides* Y9; (2) *Candida* Y1; (3) *Achromobacter* No. 5; (4) *Achromobacter* No. 483; (5) *Achromobacter* No. 7; (6) *Pseudomonas*. (Scott, 1936b).

Nutrition

The biochemical versatility and/or variability of microorganisms is such that few naturally occurring or even synthetic substances are not attacked by one species or another. Growth on some substrates, generally those which are inherently nutritionally poor or which are not readily hydrolyzed to yield metabolizable substrates, will be very slow. On nutritionally limiting substrates, the inhibitory effect of reduced a_w will usually be enhanced.

In the case of bacteria, the influence of nutrition on the water relations of growth has been demonstrated clearly with salmonellae. Comparisons of growth rates in a simple glucose–inorganic salts medium, with or without defined supplements, and a complex medium (brain–heart infusion) are shown in Fig. 5.6 (Scott, 1957). Five amino acids stimulated growth rate at optimal a_w by about 15%, while reducing the minimal a_w for growth by some 50%. Vitamin additions produced further small changes in the same directions. The complex medium greatly increased the maximal growth rate, but did not permit growth at a_w levels lower than those recorded in the supplemented synthetic medium. It follows that, while in some cases specific nutrients can be identified—in this instance, the amino acid, proline—which permit growth at lower a_w levels, there are also "luxury" nutrients that stimulate growth only under relatively favorable condi-

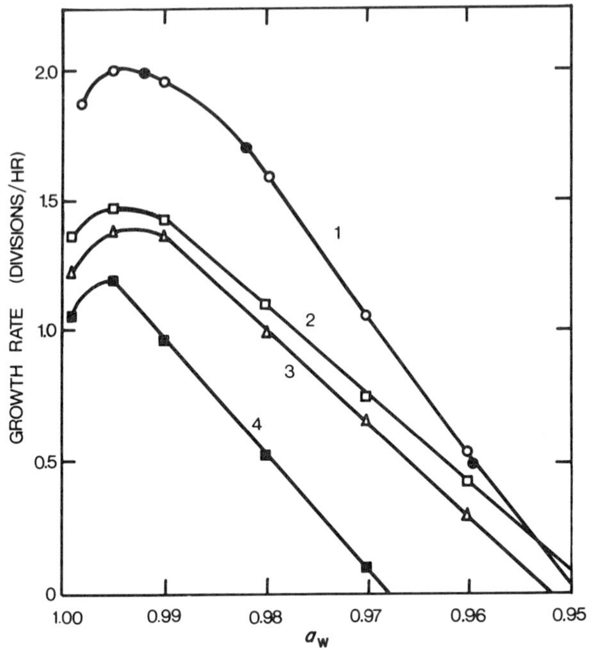

Fig. 5.6. Effect of nutrients on the rates of growth of *Salmonella oranienburg* at various a_w levels. Curve 1: brain–heart infusion broth: ●, a_w controlled by adjusting water contents; ○, a_w controlled by salts mixture. Curve 2: casein hydrolysate–yeast extract + salts mixture. Curve 3: Glucose–salts medium + five amino acids + eight vitamins + salts mixture. Curve 4: Glucose–salts medium + salts mixture (Scott, 1957).

tions and do not serve to increase the organism's tolerance to physiological drought.

The data of Snow *et al.* (1944) demonstrate the influence of nutrition on the water requirements of fungi. *Aspergillus repens* produced visible growth on bran fiber at 0.80 but not at 0.75 a_w, but growth was observed also at 0.72 a_w when starch was added to the bran. On a mixture of starch and egg albumen, this mold grew in 5 months at 0.70 a_w, but not in 7 months at this a_w on either substrate individually.

Changes in the composition of a substrate may produce subtle alterations in the environment not directly related to nutrient status. For example, altered buffering capacity could influence the ability of a microbial population to develop. Some components may alter the physical structure, leading to a change in the amount of hysteresis. This, in turn, may affect growth, as will be discussed later.

SPORULATION, GERMINATION, AND OUTGROWTH

Little attention has been paid to the effects of a_w on bacterial sporulation. In contrast, the water relations of germination and outgrowth of spores of both *Bacillus* and *Clostridium* spp. have been extensively investigated.

There is general agreement that spores may initiate germination at a_w levels appreciably lower than those which will permit outgrowth to occur. Gould (1964) showed that about twice the concentration of NaCl is required to inhibit initiation of *Bacillus* spores than is necessary to prevent outgrowth of spores in which germination has been initiated.

In *Bacillus cereus,* initiation of germination and outgrowth showed similar minimal a_w levels when a_w is controlled with salts. However, when glycerol is used, initiation at low a_w levels is not followed by outgrowth and formation of vegetative populations (Jakobsen *et al.,* 1972).

The most important spore-forming bacterium in foods is *Clostridium botulinum*. Initiation of spore germination of types A, B, and E can occur at a_w levels lower than the minima for either outgrowth or growth from vegetative inocula (Baird-Parker and Freame, 1967).

As with bacteria, the sporulation of fungi at low a_w has received little attention. As Pitt (1975) points out, it is of considerable interest, since the minimal a_w for sporulation represents the most stringent conditions in terms of water stress under which molds can complete their life cycle. Aleuriospores are produced at 0.70 a_w by *Chrysosporium* spp. and at the very low level of 0.66 a_w by *Xeromyces bisporus*. The latter organism forms ascospores at a_w levels down to 0.73.

Germination of fungal spores need not be followed by growth, especially under extreme temperature conditions or on unfavorable substrates. Thus, at 50°C, *Aspergillus niger* germinates at a_w's down to 0.78, but fails even to grow at 0.99 a_w (Ayerst, 1969). It is essential, therefore, that with both bacteria and fungi, the initiation of spore germination in systems of reduced a_w not be taken as indicating that growth of the organism concerned will ensue.

SOLUTE EFFECTS

In applying the concept of a_w to the influence of concentrated solutions on microbial growth, it is assumed that the effects observed are a function of the availability of the solvent, water, and not of the specific nature of the solute(s). In early work in which growth of molds on a range of substrates was studied as a function of equilibrium relative humidity, the nature of the major solutes in the system was never discussed, except in terms of their contributions as nutrients. In 1953. Scott showed that for several bacteria, the published data on salt and sugar

TABLE 5.2.

The Minimum a_w and Water Contents for the Growth of *Staphylococcus aureus* in Various Media at 30°C[a]

Nutrient substrate	Solutes added to control a_w	Minimum a_w	Water as dry weight (%)
Casamino acids, yeast extract, casitone	Salts mixture[b]	0.88	375
Nutrient broth	Salts mixture[b]	0.86	315
Nutrient broth	Sucrose	0.88	60
Nutrient broth	Sucrose 3.44 *M* + salts mixture[b]	0.86	75
Milk	None	0.86	16
Meat	None	0.88	23
Soup	None	0.86	63

[a] Scott, 1961.
[b] $NaCl:KCl:Na_2SO_4$::5:3:2 moles.

tolerance are equivalent in terms of a_w and went on to compare the growth response of *Staphylococcus aureus* to salts and sucrose in liquid media and to rehydrated dried foods. The most important factor limiting growth in these experiments was a_w, and not the major solute present nor the water content, which, in the various systems, at the minimal a_w for growth, ranged from 16 to 375% of the dry weight (Table 5.2) (Scott, 1961).

As would be expected, there are many solutes which are inherently toxic (e.g., salts of heavy metals), and these will exert an inhibitory effect independent from, and generally much greater than, that due to their influence on a_w. However, there also exist solutes that are much less inhibitory to the growth of some organisms than their a_w in solution might suggest. The most important of these solutes is probably glycerol, shown to be appreciably less inhibitory on an a_w basis for salmonellae than salt mixtures, glucose, or sucrose (Christian, 1955). Similar anomalous data with glycerol were reported for *Clostridium botulinum* types A, B, and E (Baird-Parker and Freame, 1967) and for *Bacillus cereus* (Jakobsen *et al.*, 1972). In the latter case, erythritol and dimethylsulfoxide were also less inhibitory than salts, but more so than glycerol.

The above data refer to bacteria with relatively high water requirements, with minimal a_w values for growth (in NaCl) in the range 0.97–0.94. A survey of the salt and glycerol tolerance of 16 species of nonhalophilic bacteria was conducted by Marshall *et al.* (1971). While the data referred to above were confirmed, the more salt-tolerant bacteria, predominantly cocci, were substantially more sensitive, on an a_w basis, to glycerol than to salt. While the minimal a_w levels for the 16 species ranged from 0.97–0.83 in NaCl-adjusted media, the range was 0.97–

0.89 a_w in glycerol. If a vibrio with a requirement for high ionic strength is deleted, the latter range becomes 0.95–0.89 a_w. These considerations are important in the context of intermediate moisture foods, in which glycerol and related substances may be used to reduce a_w.

The extremely halophilic bacteria, discussed earlier, have evolved to grow only at low levels of a_w, and only when these levels are produced by high NaCl concentrations. Surprisingly, there appear to be no reports of bacteria with a preference for environments rich in sugars.

The common nonosmophilic yeast, *Saccharomyces cerevisiae,* grows at a_w levels down to 0.93–0.92 in NaCl and to 0.91–0.90 in sucrose media. The so-called osmophilic yeasts will grow in substantially more concentrated environments and have been isolated from both sugar-rich and salt-rich foods. For many such isolates, there is surprisingly little difference between their responses to sugars and salts on an osmotic pressure (or a_w) basis (Ōnishi, 1963). Growth commonly occurs in the presence of both types of solutes at a_w levels approaching 0.80. Some isolates of *Saccharomyces rouxii,* in particular, are highly xerophilic, von Schelhorn (1950) having reported growth in fructose syrups at 0.62–0.65 a_w.

HYSTERESIS EFFECTS

The theory of hysteresis was discussed briefly in Chapter 1. It was pointed out that the a_w of a solution may be reduced, not only by hydration of solute molecules, but also by interactions with solids. The most important of these interactions in foods appear to be capillary effects. It is believed that the hysteresis loop results largely from the different energy requirements for filling and emptying capillary and similar spaces during the adsorption and desorption of water.

Labuza *et al.* (1972) have studied microbial growth in food products showing pronounced hysteresis effects. At a_w levels produced in the food by desorption, microbial responses were similar to those obtained at equivalent a_w levels in laboratory media. In contrast, when the moisture contents of the food were adjusted by adsorption, microbial growth was inhibited at substantially higher a_w values. Labuza concludes that in such foods, a_w alone is not the growth-controlling factor—both a_w and total water content are important.

This remains a controversial area, partly because quantitative experiments involving microbial growth on foods prepared by desorption and adsorption are not sufficiently reproducible, and partly because knowledge of the behavior of water in such systems is inadequate. However, from the practical point of view, it is fortunate that in adsorption-type systems, minimal a_w levels for microbial growth are higher, and not lower than predicted from earlier studies in laboratory media. Data obtained from desorption systems agree well with predictions. The hysteresis effect is likely to be of major concern in the formulation of microbiologically stable, intermediate moisture foods, as the margin of safety in terms of a_w is small.

PHYSIOLOGICAL BASIS OF TOLERANCE OF REDUCED WATER ACTIVITY

The demonstration with several halophilic and nonhalophilic bacteria that the freezing point of cells is always as low or lower than that of the growth medium (Christian and Ingram, 1957) confirmed what was obvious from thermodynamic considerations—a cell with an apparently semipermeable membrane inside a structural cell wall cannot maintain an internal environment more dilute (of higher water activity) than the external solution.

In extremely halophilic bacteria, potassium salts appear to be the major component of this high internal solute concentration. Cells grown in 4.0 *M* NaCl (0.84 a_w) medium contained some 4.5 molal potassium and 1.4 molal sodium (Christian and Waltho, 1962). Enzymes of such bacteria are clearly adapted to this situation, being much more active in high KCl concentrations than in equivalent concentrations of NaCl (Baxter and Gibbons, 1956). It seems that the ability to accumulate these high levels of potassium and to tolerate them in the cell's internal environment is the key to the salt tolerance of halophiles.

For nonhalophilic bacteria, the first solutes shown to be related to growth at reduced a_w were potassium and proline (Christian and Hall, 1972). Subsequently, Measures (1975) showed that growth of a range of such bacteria at reduced a_w involves several amino acids, those least equipped to grow at low a_w accumulating glutamic acid, more tolerant species accumulating α-aminobutyric acid and the most tolerant group, e.g., *Staphylococcus aureus,* accumulating proline.

Although fungi have lower water requirements than most bacteria, studies of the composition of molds in relation to growth at low a_w have not been reported. In contrast, osmophilic yeasts have been investigated and found to accumulate polyhydric alcohols, such as arabitol (Brown, 1974), to very high levels and to contain enzymes that are active in such situations (Brown and Simpson, 1972).

Thus, the study of the physiological basis of the microbial cell's response to reduced a_w is in its infancy. All that can be said now is that the bacteria studied to date accumulate predominantly potassium and/or amino acids, and the yeasts concentrate polyols. Molds have been virtually ignored.

REFERENCES

Ayerst, G. (1966). Influence of physical factors on deterioration by moulds. *SCI Monogr.* **23,** 14–20.

Ayerst, G. (1969). The effects of moisture and temperature on growth and spore germination in some fungi. *J. Stored Prod. Res.* **5,** 127–141.

Baird-Parker, A. C., and Freame, B. (1967). Combined effect of water activity, pH and temperature on the growth of *Clostridium botulinum* from spore and vegetative cell inocula. *J. Appl. Bacteriol.* **30,** 420–429.

Baxter, R. M., and Gibbons, N. E. (1956). Effects of sodium and potassium on certain types of *Micrococcus halodenitrificans* and *Pseudomonas salinaria*. *Can. J. Microbiol.* **2,** 599.

Brown, A. D. (1974). Microbial water relations: Features of the intracellular composition of sugar-tolerant yeasts. *J. Bacteriol.* **118,** 769–777.

Brown, A. D., and Simpson, J. R. (1972). Water relations of sugar-tolerant yeasts: The role of intracellular polyols. *J. Gen. Microbiol.* **72,** 589–591.

Christian, J. H. B. (1955). The water relations of growth and respiration of *Salmonella oranienburg* at 30°C. *Aust. J. Biol. Sci.* **8,** 490–497.

Christian, J. H. B., and Hall, J. M. (1972). Water relations of *Salmonella oranienburg:* Accumulation of potassium and amino acids during respiration. *J. Gen. Microbiol.* **70,** 497–506.

Christian, J. H. B., and Ingram, M. (1957). The freezing points of bacterial cells in relation to halophilism. *J. Gen. Microbiol.* **20,** 27–31.

Christian, J. H. B., and Scott, W. J. (1953). The water relations of salmonellae at 30°C. *Aust. J. Biol. Sci.* **6,** 565–573.

Christian, J. H. B., and Waltho, J. A. (1962). Solute concentrations within cells of halophilic and non-halophilic bacteria. *Biochim. Biophys. Acta* **65,** 506–508.

Gould, G. W. (1964). Effect of food preservatives on the growth of bacteria from spores. *Proc. Int. Symp. Food Microbiol., 4th, 1963* pp. 17–24.

Jakobsen, M., Filtenborg, O., and Bramsnaes, F. (1972). Germination and outgrowth of the bacterial spore in the presence of different solutes. *Lebensm.-Will. u. Technol.* **5,** 159–162.

Labuza, T. P., Cassil, S., and Sinskey, A. J. (1972). Stability of intermediate moisture foods. 2. Microbiology. *J. Food Sci.* **37,** 160–162.

Marshall, B. J., Ohye, D. F., and Christian, J. H. B. (1971). Tolerance of bacteria to high concentrations of NaCl and glycerol in the growth medium. *Appl. Microbiol.* **21,** 363–364.

Measures, J. C. (1975). Role of amino acids in osmoregulation of non-halophilic bacteria. *Nature (London),* **257,** 398–400.

Ohye, D. F., and Christian, J. H. B. (1967). Combined effects of temperature, pH and water activity on growth and toxin production by *Clostridium botulinum* types A, B and E. *Botulism 1966, Proc. Int. Symp. Food Microbiol., 5th, 1966* pp. 217–223.

Ōnishi, H. (1963). Osmophilic yeasts. *Adv. Food Res.* **12,** 53–94.

Pitt, J. I. (1975). Xerophilic fungi and the spoilage of foods of plant origin. *In* "Water Relations of Foods" (R. B. Duckworth, ed.), pp. 273–307. Academic Press, New York.

Scott, W. J. (1936a). The growth of microorganisms on ox muscle. I. The influence of water content of substrate on rate of growth at −1°C. *J. Counc. Sci. Ind. Res. (Aust.)* **9,** 177–190.

Scott, W. J. (1936b). The growth of microorganisms on ox muscle. III. The influence of 10 per cent carbon dioxide on rates of growth at −1°C. *J. Counc. Sci. Ind. Res. (Aust.)* **11,** 266–277.

Scott, W. J. (1953). Water relations of *Staphylococcus aureus* at 30°C. *Aust. J. Biol. Sci.* **6,** 549–564.

Scott, W. J. (1957). Water relations of food spoilage microorganisms. *Adv. Food Res.* **7,** 83–127.

Scott, W. J. (1961). Available water and microbial growth. *Proc. Low Temp. Microbiol. Symp., 1961* pp. 89–105.

Snow, D., Crichton, M. H. G., and Wright, N. C. (1944). Mold deterioration of feeding stuffs in relation to humidity of storage. *Ann. Appl. Biol.* **32,** 102–110 and 111–116.

Troller, J. A. (1971). Effect of water activity on enterotoxin B production and growth of *Staphylococcus aureus. Appl. Microbiol.* **21,** 435–439.

von Schelhorn, M. (1950). Untersuchungen über den Verberb wasserarmer Lebensmittel durch osmophile Mikroorganismen. I. Verberb von Lebensmittel durch osmophile Hefen. *Z. Lebensm.-Unters.-Forsch.* **91,** 117–124.

Wodzinski, R. J., and Frazier, W. C. (1960). Moisture requirements of bacteria. I. Influence of temperature and pH on requirements of *Pseudomonas fluorescens. J. Bacteriol.* **79,** 572–578.

Wodzinski, R. J., and Frazier, W. C. (1961a). Moisture requirements of bacteria. II. Influence of temperature, pH, and malate concentration on requirements of *Aerobacter aerogenes. J. Bacteriol.* **81,** 353–358.

Wodzinski, R. J., and Frazier, W. C. (1961b). Moisture requirements of bacteria. III. Influence of temperature, pH, and malate and thiamine concentration on requirements of *Lactobacillus viridescens. J. Bacteriol.* **81,** 359–365.

Wodzinski, R. J., and Frazier, W. C. (1961c). Moisture requirements of bacteria. IV. Influence of temperature and increased partial pressure of carbon dioxide on requirements of three species of bacteria. *J. Bacteriol.* **81,** 401–408.

Wodzinski, R. J., and Frazier, W. C. (1961d). Moisture requirements of bacteria. V. Influence of temperature and decreased partial pressure of oxygen on requirements of three species of bacteria. *J. Bacteriol.* **81,** 409–415.

6

Food Preservation and Spoilage

Many foods are preserved, particularly from microbial attack, by control or reduction of the level of a_w. This chapter considers the preservation of foods in this category, the role that microorganisms play in preserving some of them and the spoilage patterns that ensue as a consequence of inadequate processing or storage. The problems of maintaining desired levels of a_w are discussed in Chapter 10.

CEREALS AND LEGUMES

Although the seeds of crops in this category lose moisture as they mature, some are harvested at high water contents, e.g., peanuts with levels of a_w in excess of 0.90. All are subject to some form of mold attack if the a_w is not reduced sufficiently rapidly to a safe level. The molds which cause damage, the storage fungi, are capable of invading the grains and kernels at minimal a_w levels that range from 0.85 for some species to less than 0.70 for others. They are, therefore, by the definition of Pitt (1975), xerophilic.

Only a few genera of molds contain xerophilic species—*Aspergillus, Eurotium* (also known as the *Aspergillus glaucus* group), *Chrysosporium, Emericella* (the *Aspergillus nidulans* group), *Eremascus, Xeromyces, Paecilomyces, Penicillium,* and *Wallemia* (*Sporendonema*). Of these, species of *Eurotium* and *Aspergillus* have been most frequently incriminated as spoilage fungi of cereals (Christensen and Kaufmann, 1965). These groups have lower a_w limits for growth of close to 0.70 and 0.75, respectively. However, Pitt (1975) points out that isolation methods have not been conducive to detecting such likely candidates as *Eremascus, Chrysosporium,* and *Xeromyces*.

Concern about the mold spoilage of stored commodities has increased with the realization that some of the species involved are toxigenic. The most xerophilic of the toxigenic molds is believed to be *Aspergillus ochraceus,* whose minimal a_w for growth is below 0.76 and for production of penicillic acid and ochratoxin A has values of about 0.80 and 0.85, repectively (Bacon *et al.,* 1973). While this species is relatively common on cereals, few strains appear to be toxigenic.

The toxin of greatest concern in peanuts is aflatoxin, produced by *Aspergillus flavus* and the closely related *Aspergillus parasiticus*. Although *A. flavus* grows at a_w levels down to 0.78, it fails to produce aflatoxin in inoculated, mature peanut kernels incubated at 83% R.H. for 84 days (Diener and Davis, 1970).

"Safe," "limiting," or "alarm" water contents have been recommended for many dry foods and are in wide use commercially, especially for the commodities discussed in this section. As Scott (1957) has demonstrated, this practice is often unwise. The slope of many water sorption isotherms is relatively flat in the range 0.6–0.8 a_w, and, for some foods, changes in water content of only 1% may result in changes in a_w of 0.04–0.08. In some products, the estimation of

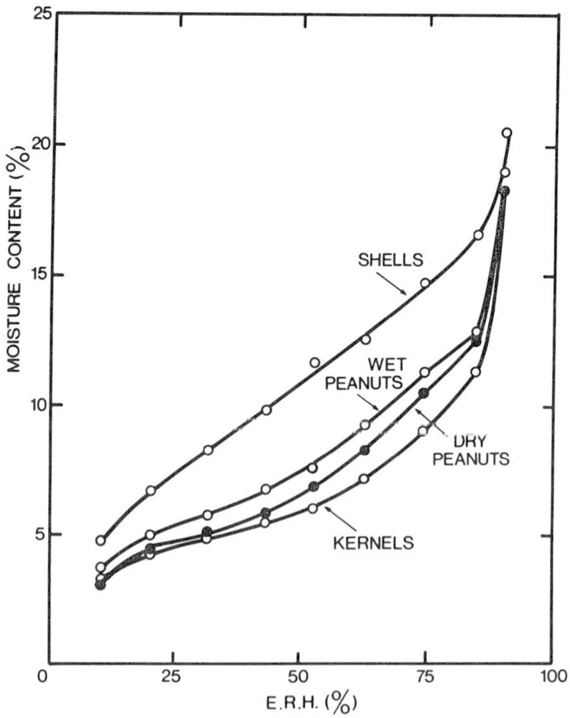

Fig. 6.1. Water sorption isotherms of whole peanuts, shells, and kernels of naturally cured runner peanuts (Karon and Hillery, 1949).

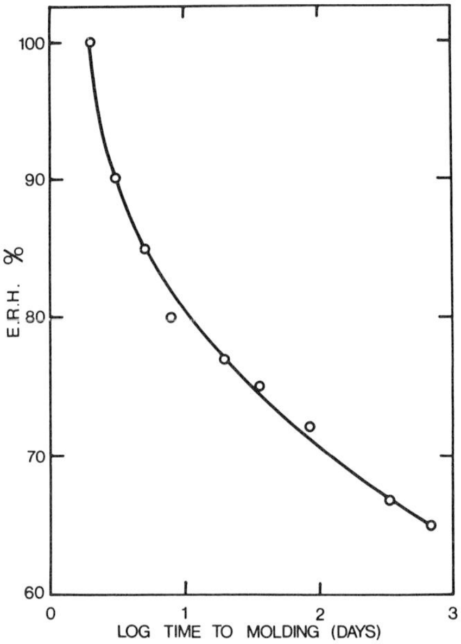

Fig. 6.2. Effect of a_w on the rate of molding of locust beans (Snow *et al.*, 1944).

water content is difficult, and errors of this magnitude can arise. Such changes can have important consequences in terms of storage life. In addition, the isotherms of different samples of a material, or of different varieties of a crop, may differ appreciably.

Confusion can arise when the form of a commodity is not specified. An example is peanuts, for which safe water contents have been suggested for the whole peanut and for the kernel alone. Figure 6.1 shows clearly the differences in water content of these two commodities at the same a_w in the 0.50–0.75 range. It is for reasons such as these that a_w is considered to be preferable to water content as an indicator of the microbiological stability of a food.

The first demonstration of the relationship between a_w and storage life was by Snow *et al.* (1944). Studies of the time taken for visible mold to develop at 21°–27°C on a range of foodstuffs held at various humidities led to the type of curve shown in the semilogarithmic plot of Fig. 6.2. The curve reproduced is for locust beans, chosen because this commodity proved the most susceptible to attack at all a_w levels studied.

Isotherms subsequently prepared for other foods have shown this curve to be a useful guide to the storage life of foods held at moderate temperatures and preserved only by reduced a_w. Foods at lower temperatures, of low pH, or containing chemical preservatives will store longer at a given a_w. Conversely, those held at higher (e.g., tropical) temperatures would be less stable to microbial attack than predicted by this curve. Accepting all of these qualifications, the data of Fig. 6.2 indicate a storage life of about 1 month at 0.75 a_w, 5 months at 0.70 a_w, and 2 years at an a_w not exceeding 0.65. *Xeromyces bisporus and Eurotium echinulatum* are the only molds that have been reliably reported as growing below 0.65 a_w.

While it is clearly wise to allow a safety margin when setting a maximal a_w level for storage of a food, excessive dehydration should be avoided. Both cereals and legume seeds become brittle and tend to shatter, which not only reduces quality directly but also increases vulnerability to insect attack. Rice is said to lose flavor if dried below 15% moisture, and grains to be milled will be subsequently "conditioned" by moistening. In addition, there is the economic penalty imposed by loss of salable weight.

FISH

Because of its highly perishable nature, fish is traditionally preserved in many cultures by methods that reduce a_w. These techniques include salting, drying, smoking, and pickling. Although production of dried and cured fish is decreasing in Western countries, it seems likely that developing nations in tropical areas may look increasingly to refining these processes in the future.

Salting is most commonly carried out by stacking suitably prepared fish in alternate layers with dry salt (see also Chapter 9). The brine produced by the extraction of water from the fish may be drained away to give a dry salted product or retained to build up over several days until the fish is covered. Subsequently, the fish may be sun-dried or processed in artificial driers. Smoking may also be carried out. If salt penetrates, moisture exudes sufficiently rapidly, and more salt is added to maintain a near-saturated brine, the product will be unspoiled when salting is complete.

The spoilage bacteria of heavily salted fish are classically the red obligate halophiles *(Halobacterium* and *Halococcus),* which produce pink discoloration and are strongly proteolytic. They are, however, aerobic and also are unable to multiply at temperatures below about 10°C. Thus, salted fish, both wet stacked and after subsequent drying, are exposed to air, when susceptible to "pink" spoilage at ambient temperatures, but are readily protected by refrigeration.

Even with heavily salted fish (about 0.75 a_w), microbiological putrefaction may become evident in 2–3 months at temperatures of about 20°C. Below 10°C, bacterial spoilage may be prevented for years, although the fish may become so soft and rancid as to be inedible. More lightly smoked fish (about 0.85 a_w) should be stored at 0°C or lower.

Dried fish is also subject to attack by the "dun" mold, *Wallemia sebi,* which forms chocolate-brown spots. It does not however cause objectionable flavor or texture changes in the fish and can be controlled by dipping fish in a solution of 0.1% sorbic acid.

Fish may be preserved from microbiological attack in concentrated brines for long periods, with the qualification that autolytic changes will proceed. Changes will be particularly rapid if fish are not gutted. Such a process is used to produce fish pastes and sauces. Those cured in weaker brines, particularly brines containing sucrose, and those transferred to sugar-containing sauces may be spoiled by the production of "rope." This results from macromolecular slimes formed by bacteria of the genera *Achromobacter* and *Leuconostoc*. The rope may consist of polysaccharides or nitrogen-containing polymers. The best studied is a laevan produced by a moderately halophilic *Achromobacter* sp. (Lindeberg, 1957).

Salting may be followed by smoking, either "cold" (20°–45°C) or "hot" (65°–90°C). The keeping quality of smoked fish results partly from the antimicrobial action of certain smoke constituents, but mostly from the salting and drying they have received. Such fish may indeed be very lightly salted, as those implicated in an outbreak of type E botulism in the United States (Anonymous, 1964), although undercooking, vacuum packaging, and poor sanitation were probably more important factors. *Clostridium botulinum* type E does not appear to grow at a_w levels below 0.965, corresponding to 5–6% NaCl in the water phase. Fish salted to this level, even if smoked, will require refrigeration if growth of many spoilage bacteria, particularly micrococci, is to be controlled. On the other hand, heavily salted, heavily smoked fish are generally resistant to bacterial attack, but may be subject to mold deterioration, as are the dried products discussed earlier.

MEAT

Water activity may play an important role in the long-term chilled storage of carcass meats. The growth of psychrotrophic bacteria, predominantly *Pseudomonas* spp., causing spoilage of fresh meats, is slowed significantly, particularly during the initial chilling period, by the reduction in a_w that results from surface drying (Scott and Vickery, 1939).

The preparation of meat extracts involves dehydration under vacuum at elevated temperatures. The resulting paste is microbiologically stable by virtue of its high salt concentration and, hence, low a_w. However, thermophilic bacteria grow rapidly during early stages of dehydration and will cause spoilage unless the initial spore load is low. The water relations of growth of a thermophilic bacterium, *Bacillus subtilis,* at 50°C are depicted in Fig. 6.3. Note that in salt broth, this organism can divide hourly at 0.93 a_w.

Of great significance is the preservation of meats by curing. In cured pork

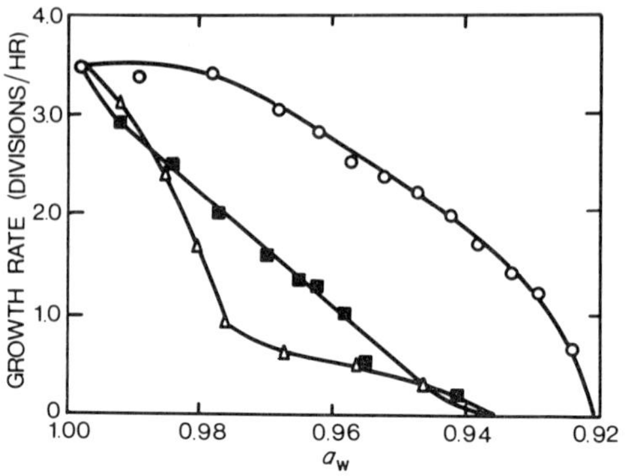

Fig. 6.3. Water relations of growth of *Bacillus subtilis* at 50°C. The basal medium was brain-heart infusion, adjusted to required levels of a_w by addition of sucrose (△), sodium chloride (○), or glycerol (■).

(bacon, ham), beef, mutton, and other meats and in cured sausage, reduction in a_w by addition of salt is the predominant preserving method, but not always the only important one. Nitrite and sometimes nitrate are also involved in the cure, and some products are also smoked. Fermented sausages contain acids and other preservative products of microbial metabolism. National variations exist in the formulating and processing of cured meats, and so valid generalizations are difficult to make.

Nitrite plays several roles in the curing of meat. Most obviously, it combines with the myoglobin of fresh muscle to form the red pigment, nitrosomyoglobin, which is converted on heating to the desirable pink nitrosohemochrome. It also has some inhibitory effect at the pH of cured meats on the growth of some spoilage bacteria. However, its most important role microbiologically is in the control of *Clostridium botulinum* in canned, cured meats. Growth of *C. botulinum,* types A and B, the types most likely to cause botulism from meat products, is prevented at NaCl concentrations of 8 and 10%, respectively, i.e., at a_w levels below 0.95 and 0.94 (Ohye and Christian, 1967). These levels of salt are now organoleptically unacceptable in most cured meats, and the levels in canned cured products usually do not exceed 6%. The safety of such products after moderate heating stems from the ability of nitrite or some derivative to interfere with the germination or outgrowth of surviving heat-treated spores (Roberts and Ingram, 1966; Perigo *et al.,* 1967).

Growth of the predominant aerobic spoilage agents of fresh meat, the gram-negative, rod-shaped bacteria, is substantially inhibited in even the more lightly cured meats, e.g., frankfurters, bologna, and liver sausages, which have a_w levels near 0.97 (5% NaCl). Spoilage of these cooked products is commonly by members of the Lactobacillaceae, producing acid (souring) and converting the red, modified meat pigment to green. Greening bacteria include *Lactobacillus viridescens* and *Streptococcus faecium*. Classification of this group of bacteria remains confused (Kitchell and Shaw, 1975). Control of spoilage is achieved by a cooking process sufficient to destroy most organisms of this type, and also by storage below 5°C.

Normal smoking of frankfurters lowers the bacterial load appreciably and may leave micrococci predominant. These will grow during chilled storage. The fermented sausages, which may also be smoked, depend for their stability on the lowered pH produced by the lactic fermentation and, in the case of the dry sausages, on a lower a_w than is found among the sausages referred to above. The final a_w is influenced by the period of ripening as well as by the formulation. Leistner and Rödel (1975) list the minimal, maximal, and modal a_w values for fermented sausage as 0.72, 0.95, and 0.91 and quote, as an example, Hungarian salami with an a_w value of about 0.83. Lactic acid bacteria, such as *Pediococcus cerevisiae,* have been advocated as pure culture starters for fermented sausage.

Dry sausages of the Hungarian or Italian type are frequently coated with mold, most commonly, *Penicillium, Scopulariopsis,* or *Aspergillus* spp. (Leistner and Ayres, 1968). Providing the mold forms only a thin cover, is not toxigenic, and is light in color, it is generally considered desirable on such products and may contribute to their flavor. Mold starter cultures, such as *Penicillium nalgiovensis,* have been recommended for the ripening of fermented sausages (Mintzlaff and Christ, 1973).

Although many micrococci are most salt tolerant than are lactobacilli, they are sensitive to pH in the range 5–6. Hence, in a salted product that contains fermentable carbohydrate, lactobacilli will frequently predominate by virtue of the acid they produce.

Pork and beef cuts, whether cured by dry ingredients, by pumping with pickle, by immersing in brine, or by some combination of these methods, present similar microbiological pictures. If, as in some traditional processes, the cure contains nitrate, but no nitrite, bacteria are necessary to reduce the former to the latter, and this appears to be a role for certain micrococci. The brines, containing 15% or more NaCl (0.85 a_w) (generally $<$ 20% in the case of hams and $>>$ 20% for bacon), develop high numbers of undesirable halophilic bacteria. *Micrococcus halodenitrificans,* which can grow in $>$ 20% NaCl ($<$ 0.865 a_w), depletes the nitrite levels in the cure and so contributes to the instability of the finished product.

The microflora of aerobically stored bacon consists of salt-tolerant micrococci and gram-negative rods. When sliced and vacuum packed, the predominant organisms continue to be micrococci if the salt concentration is high (8–12% in the aqueous phase, 0.95–0.92 a_w), but these are succeeded by lactic acid bacteria in low-salt bacon (5–7% in the aqueous phase, 0.97–0.95 a_w) (Tonge *et al.*, 1964).

TABLE 6.1.

Storage Categories of Meat Products Based on the a_w and pH Values of the Product, with Recommended Storage Temperatures[a]

Category	Criteria	Temperature
Easily perishable	pH > 5.2 and a_w > 0.95	≤ +5°C
Perishable	pH 5.2–5.0 (inclusive) *or* a_w 0.95–0.91 (inclusive)	≤ +10°C
Shelf stable	pH ≤ 5.2 *and* a_w ≤ 0.95 *or* only pH < 5.0 *or* only a_w < 0.91	No refrigeration required

[a] Leistner and Rödel, 1975.

As with fish, meat and animal by-products preserved by heavy salting will be prone to proteolytic spoilage by obligately halophilic bacteria, commonly the "red halophiles," although similar organisms with no or different pigmentation may be isolated. Such spoilage occurs only if the product is held under aerobic conditions at temperatures above 10°C. It is found in dry-salted, natural sausage casings (animal intestines), especially if inadequately cleaned, and in pickled pork and beef meats packed in saturated brines.

Leistner and Rödel (1975) offer a useful differentiation of meat products into three storage categories based on their a_w and pH values (Table 6.1). It should be noted, however, that these are traditional products, and in some, intrinsic factors other than a_w and pH influence storage life. Other foods at similar levels of a_w and pH may not possess the same stability as the products listed here.

MILK PRODUCTS

Many of the microaerophilic lactic acid bacteria responsible for flavor development and ripening of cheese and other dairy products are relatively osmotolerant. Optimal growth rates commonly are observed at a_w levels of 0.95 and above. Growth minima are in the 0.92–0.90 range. Lactic acid synthesis occurs at 1.0–0.95 a_w but declines sharply below this level commensurate with growth inhibition. Production of flavor compounds, such as diacetyl (2,3-butanedione) frequently is optimal at 0.97 a_w (J. A. Troller, personal data).

The main milk products preserved by control of a_w level are dried whole and nonfat milk. With normal a_w levels of about 0.2, these are subject to microbial spoilage only when gross rehydration occurs.

Sweetened condensed milk, with a final sugar concentration of 55–60% and an a_w level of about 0.85–0.89, is not heat treated after the cans have been filled and sealed, but experiences mild heating prior to and during concentration. This

TABLE 6.2.

Water Activity Levels of Several Cheese Varieties Obtained from Retail Markets in the United States[a]

Cheese variety	a_w
Process cheddar	0.93
Natural cheddar, longhorn style	0.93
Muenster	0.94
Swiss	0.94
Provolone	0.92

[a] Personal data of the authors.

effectively pasteurizes the milk, eliminating vegetative bacteria, molds, and yeasts, but recontamination may occur.

If air is present in the headspace, mold "buttons" may develop. Gas formation, particularly by the non-lactose-fermenting yeast *Torulopsis lactis-condensi*, is another spoilage problem. These microbial activities take place at a_w levels in the 0.90–0.85 range.

Cheeses vary greatly in their moisture content, and consequently in their a_w (Table 6.2) and in the types of spoilages that may occur. The a_w level plays little part in early stages of production or preservation of cheese. After coagulation and cutting of the curd, expulsion of whey reduces water content, but the whey carries with it much of the lactose and salts that would reduce the a_w in the residual aqueous phase in the curd. Subsequent salt addition gives a salt concentration in the finished cheese of, for example, 1.5–2.0% for cheddar and 3.4–4.5% for blue-vein cheeses. Final levels of a_w may be 0.93 for processed cheddar cheese, 0.92 for a cheese of medium age, and around 0.89 for an old cheese. Clearly all three types of cheese will be susceptible to surface mold spoilage, the first being protected by an airtight foil wrapping and refrigeration, the second by refrigeration and sometimes moisture-proof wrapping or waxing, and the third by the hard "skin" developed on drying or by waxing. The genera listed earlier as growing on dry sausage, *Penicillium, Scopulariopsis,* and *Aspergillus,* contain species that also grow on cheese surfaces, and *Monilia nigra* forms black spots on the rind of hard cheese.

VEGETABLES

Fresh vegetables require controlled temperatures and relative humidities to minimize moisture loss and spoilage during extended storage. The conditions considered appropriate for the storage of a range of vegetables are discussed in Chapter 10 and listed in Table 10.1 of that chapter.

The dehydration processes applied to many vegetables, such as potatoes, onions, peas, and beans, result in products of some 5–10% residual water and a_w levels in the range 0.10–0.35. Such foods run no risk of spoilage unless extensive hydration occurs, and so the products must be packaged in containers that prevent rehydration (see Chapter 10).

Vegetables for dehydration arrive at the processing plant heavily contaminated with microorganisms. Lye peeling of tomatoes and some root vegetables reduces the surface count substantially. An efficient blanching or cooking decreases the load by more than 99.9%, leaving bacterial spores as virtually the only living organisms. Vegetables such as onions, which are not blanched, can enter the dehydrator very heavily contaminated. Vaughn (1951) described the rapid microbial growth and spoilage of vegetables that ensues when a tray or belt is loaded in a way that impedes heat and mass transfer. The flora of dried onions spoiled in this way was predominantly *Lactobacillus* spp. and that of potatoes was mostly the starch-fermenting species of *Aerobacter*. In other methods of drying—by spray, drum, or roller—there is much less opportunity for this type of spoilage, but any preheating or holding step that is poorly controlled can lead to similar problems.

Microorganisms play important roles in the preservation by pickling of such vegetables as cabbage, olives, and cucumbers. The processes are based on salting which, by reducing the a_w level, restricts the growth of gram-negative bacteria and permits the growth and lactic acid production by naturally contaminating members of the *Lactobacteriaceae* (Frazier, 1967). The reduced pH controls growth of salt-tolerant micrococci and depletion of fermentable carbohydrate limits the development of salt- and acid-tolerant spoilage bacteria and yeasts. Film-forming yeasts may pave the way for colonization of the surface by other microorganisms and even by insects.

Oriental cultures have many fermented vegetable foods based not only on cabbage but also on starchy tissues, such as Hawaiian poi from taro roots. Between these extremes are mixed fermentations, such as Korean tongkimchi, in which ground fish, meat, nuts, or rice water is combined with the cabbage. The role of a_w in most of these processes is not clear, although salt is involved in many of them. Microbiologically, most is known of the fermentations of cabbage, cucumbers, and olives (Pederson, 1960).

Sauerkraut is produced by brining shredded cabbage under anaerobic conditions. The ideal NaCl concentration appears to be close to 2% (0.988 a_w). This is adequate to retard the growth of gram-negative bacteria, but low enough to permit rapid growth of *Leuconostoc mesenteroides,* the most salt sensitive of the lactic acid bacteria concerned and the first of a sequence of several organisms to be involved in sauerkraut production. The species which follow are *Lactobacillus plantarum* and *Lactobacillus brevis,* with occasional involvement of *Pediococcus cerevisiae* and *Streptococcus faecalis* (Penderson, 1960).

Variations in salt concentration (a_w) and in temperature alter the character of the sauerkraut by changing the acidic contributions of the fermenting bacteria. The homofermentative species, *L. plantarum* and *P. cerevisiae,* produce almost entirely lactic acid, while the heterofermentors, *L. brevis* and *L. mesenteroides,* yield substantial amounts of acetic acid as well. Pure culture fermentations have shown how the production of these acids is reflected in the sauerkraut flavor (Stamer, 1968). Low temperatures (13°–18°C) and low quantities of salt (1.8–2.25%, 0.990–0.987 a_w), which favor growth of heterofermentative lactic acid bacteria, but which also permit the development of homofermentations, are said to give the highest quality sauerkraut.

Some microbiological faults of sauerkraut result from inappropriate levels of a_w. Excessive salt may encourage yeast growth, leading in some cases to pink discolorations caused by aerobic species, while too little salt permits growth of gram-negative bacteria with subsequent spoilage.

The pickling of cucumbers is similar to that of cabbage. However, higher initial salt concentrations (5–8%, 0.97–0.95 a_w) largely preclude any contribution to acid production by *Leuconostoc mesenteroides*. Acid production by other species is maximal after about a week, and the a_w is progressively reduced by salt additions to give a holding brine of about 0.87 a_w (about 17% NaCl) (Etchells *et al.*, 1975).

Undersalting (below 5%, 0.97 a_w) may permit coliform bacteria to become established if large numbers of lactic acid bacteria are not present and able to produce acid rapidly. Oversalting encourages the growth of halophilic coliforms, and both types of coliform bacteria produce gas, particularly carbon dioxide, in amounts that cause bloated or hollow cucumbers. Oversalting may permit the growth of osmophilic yeasts, which also produce very large quantities of carbon dioxide. Finished sweet pickles, which may contain 20–40% sugar, are also susceptible to fermentation by osmophilic yeasts, especially *Saccharomyces rouxii*.

At all salt concentrations, film yeasts may occur on the brine surface. They are, by definition, aerobic, and are found among both the false or asporogenous yeasts, *Mycoderma* and *Candida,* and the true yeasts, *Pichia, Hansenula,* and *Debaryomyces*. Species of *Debaryomyces* are very tolerant of reduced a_w levels, forming films on near-saturated brines. Film yeasts oxidize lactic acid, increasing pH to levels at which growth of coliforms and lactic acid bacteria may resume (Etchells *et al.*, 1975). Their growth is controlled by direct sunlight, by a close fitting cover that excludes air from the brine tank, or by certain antimycotics.

The fermentation of olives has much in common with the pickling of cucumbers (Vaughn, 1975). The levels of a_w (0.97–0.95) are similar, as are the lactic acid bacteria involved, except that *Leuconostoc mesenteroides* appears to play a more important role in olives than in cucumbers. The major difference is in the need for lye treatment to hydrolyze the bitter glucoside, oleuropein, prior to fermentation. A major potential microbial spoilage problem is, again, growth of oxidative film yeasts, controlled as noted before by the exclusion of air.

When acid production is inadequate and the pH does not fall below 4.0, there is a risk of bacterial spoilage of the olives by species of *Propionibacterium* or *Clostridium*. These organisms are capable of converting the brine acids to propionic and higher volatile acids producing foul odors in the affected brines (Vaughn, 1975).

A number of other types of vegetables, such as green tomatoes, string beans, and corn, may be preserved by similar lactic fermentations. However, vegetables that soften readily are more commonly salted in concentrated brines (18% to saturated NaCl, 0.86–0.75 a_w), which prevent the lactic fermentation altogether. Those vegetables, such as some legumes, with available carbohydrate contents considered too low to support an adequate lactic fermentation, may also be salted to these high levels.

Tomato sauce and catsup, preserved by virtue of their high acid and solids content, may undergo fermentative spoilage by yeasts. When preserved with acetic acid at its high level of a_w, spoilage by the preservative-resistant yeasts, *Saccharomyces bailii, Pichia membranifaciens,* and *Candida krusei* has been observed (Pitt and Richardson, 1973).

Soybeans are the basis of a wide range of fermented products prepared in the Orient, among which the most important are soy sauce and miso paste. In the preparation of soy sauce, soybean mashes are cooked and inoculated with a starter culture in which the mold *Aspergillus oryzae* predominates. When a heavy growth has developed, the mash is added to a strong brine to give a final NaCl concentration of about 18% (0.86 a_w). The mold enzymes continue to act on the mash, and salt-tolerant bacteria, such as pediococci, commence a secondary fermentation, reducing the pH from about 6.0 to below 5.0 (Ōnishi, 1963). This permits rapid growth of osmophilic yeasts, particularly *Saccharomyces rouxii,* which are important contributors to the flavor of the finished product. The raw sauce, when pressed from the mash, may become infected with film yeasts, but it is claimed that mature, good quality soy sauce is protected from such spoilage by antimicrobial compounds that develop on storage.

Miso is produced from a cooked soybean paste containing from 7 to 20% NaCl (0.96–0.84 a_w) in the aqueous phase (Ōnishi, 1963). Again, the fermenting mold is *Aspergillus oryzae*. This is a relatively rapid process, and spoilage is avoided if the paste contains less than 45% water.

FRUIT

Many types of fruit are preserved by dehydration to a_w levels of 0.65–0.60. The lye peeling or alkali dipping received by some substantially reduces the surface microbial load, but for many fruits, sulfuring is of importance for long-term preservation. Light-colored fruits receive treatments that leave residual sulfur dioxide levels of 1000–3000 ppm to control nonenzymatic browning. Such fruits are unlikely to undergo microbial spoilage even at relatively high a_w levels. On the other hand, those fruits not sulfited, notably prunes and raisins, are subject to spoilage unless a_w remains low or another preservative, such as sorbic acid, is used.

Dried prunes of a_w 0.68–0.60 rarely spoil unless some rehydration occurs, in which case species of *Eurotium* (the *Aspergillus glaucus* group) are the most common fungi isolated. High-moisture prunes, containing about 35% moisture (0.94 a_w), are intended for consumption without rehydration and are highly perishable unless heated in a hermetically sealed container or treated with a suitable antimycotic. In the absence of such treatment, *Eurotium* spp. predominate among the spoilage flora. However, *Xeromyces bisporus,* with a minimum a_w for germination of 0.605, is the most frequently isolated species of spoilage organism of high-moisture prunes (Pitt and Christian, 1968). The storage of fresh fruit is discussed in Chapter 10 and the desirable conditions listed in Table 10.1 of that chapter.

The preservation of fruit in the form of jams and jellies relies on attaining a_w levels of about 0.80 or lower. This is achieved by partial inversion of sucrose during cooking, which leaves a product free of vegetative fungi and microbiologically stable, if protected from comtamination. However, contamination by molds commonly results in surface growth.

Liquid fruit products, such as fruit concentrates, juice concentrates, cordials, and syrups, are protected from microbial attack by combinations of low a_w, low pH, and, in some cases, added preservatives. *Saccharomyces rouxii* grew in a pear concentrate of pH 4.5 at 0.62 a_w, but not at 0.60 (von Schelhorn, 1950). This contained no preservative, and such products will commonly be spoiled by this yeast.

Preservatives such as acetic, sorbic, or benzoic acids should control *S. rouxii,* but, frequently, they do not control the very preservative-resistant species *Saccharomyces bailii,* which is a common cause of spoilage of preserved, fruit-based products of a_w greater than 0.80 (Pitt and Richardson, 1973).

Although sugar in its crystalline form is stable, spoilage due to thermophilic *Bacillus* spp. and molds can occur during evaporation of the syrup prior to crystallization. In producing honey, the sucrose is inverted by bee enzyme and, consequently, much lower a_w levels are obtained. The actual a_w of honey (Appendix A) depends upon atmospheric humidity during collection, but honey is most commonly spoiled by stains of yeasts now included in *S. rouxii* and growing at a_w levels down to 0.62. Crystallized fruit held at high humidity will also support growth of these organisms.

Canned fruit, being acid, is preserved by a relatively mild heat process. Most attention has been paid in the past to the influence of pH on the processing required. However, Jakobsen and Jensen (1975) have shown that a_w plays an important role in the stability of canned pears. Growth of the butyric anaerobe, *Clostridium pasteurianum,* which in the spore state may survive the heating given to such products, occurred at pH 3.8 if the a_w was $\geq$ 0.985, but not at pH 3.8–4.0 when the a_w was in the 0.985–0.975 range. This suggests that a shorter heat process and enhanced quality may result from the modification of a_w as well as pH in canned fruit.

CONFECTIONERY

Chocolates, toffees, caramels, and other confectioneries are protected from microbial spoilage by a_w levels that are commonly in the 0.65–0.50 range. A major spoilage problem is encountered in soft-centered chocolates which, if the fondant is predominantly sucrose, may support a yeast fermentation. The ensuing carbon dioxide production may burst the chocolate casing. This is avoided by adding invertase to the fondant, which inverts the sucrose and provides a sufficiently low a_w to inhibit the yeasts.

REFERENCES

Anonymous. (1964). Botulism outbreak from smoked whitefish. *Food Technol. (Chicago)* **18,** 71.

Bacon, C. W., Sweeney, J. G., Robbins, J. D., and Burdick, D. (1973). Production of penicillic acid and ochratoxin A on poultry feed by *Aspergillus ochraceus:* Temperature and moisture requirements. *Appl. Microbiol.* **26,** 155–160.

Christensen, C. M., and Kaufmann, H. H. (1965). Deterioration of stored grains by fungi. *Annu. Rev. Phytopathol.* **3,** 69–84.

Diener, U. L., and Davis, N. D. (1970). Limiting temperature and relative humidity for aflatoxin production by *Aspergillus flavus* in stored peanuts. *J. Am. Oil Chem. Soc.* **47,** 347–351.

Etchells, J. L., Fleming, H. P., and Bell, T. A. (1975). Factors influencing the growth of lactic acid bacteria during the fermentation of brined cucumbers. *In* "Water Relations of Foods" (R. B. Duckworth, ed.), pp. 281–305. Academic Press, New York.

Frazier, W. C. (1967). "Food Microbiology," 2nd ed. McGraw-Hill, New York.

Jakobsen, M., and Jensen, H. C. (1975). Combined effect of water activity and pH on the growth of butyric anaerobes in canned pears. *Lebensm.-Wiss. u. Technol.* **8,** 158–160.

Karon, M. L., and Hillery, B. E. (1949). Hygroscopic equilibration of peanuts. *J. Am. Oil Chem. Soc.* **26,** 16–19.

Kitchell, A. G., and Shaw, B. G. (1975). Lactic acid bacteria in fresh and cured meat. *In* "Lactic Acid Bacteria in Beverages and Food" (J. G. Carr, C. V. Cutting, and G. C. Whiting, eds.), pp. 209–220. Academic Press, New York.

Leistner, L., and Ayres, J. C. (1968). Molds and meats. *Fleischwirtschaft* **48,** 62–65.

Leistner, L., and Rödel, W. (1975). The significance of water activity for microorganisms in meats. *In* "Water Relations of Foods" (R. B. Duckworth, ed.), pp. 309–323. Academic Press, New York.

Lindeberg, G. (1957). Laevan-forming halophilic bacteria. *Nature (London)* **180,** 1141.

Mintzlaff, H.-J., and Christ, W. (1973). *Penicillium nalgiovensis* als Starterkultur für "Sudtiroler Bauernspeck." *Fleischwirtschaft* **53,** 864–867.

Ohye, D. F., and Christian, J. H. B. (1967). Combined effects of temperature, pH and water activity on growth and toxin production by *Clostridium botulinum* types A, B and E. *Botulism 1966, Proc. Int. Symp. Food Microbiol., 5th, 1966* pp. 217–222.

Ōnishi, H. (1963). Osmophilic yeasts. *Adv. Food Res.* **12,** 53–94.

Pederson, C. S. (1960). Sauerkraut. *Adv. Food Res.* **10,** 233–291.

Perigo, J. A., Whiting, E., and Bashford, T. E. (1967). Observations on the inhibition of vegetative cells of *Clostridium sporogenes* by nitrite which has been autoclaved in a laboratory medium, discussed in the context of sub-lethally processed cured meats. *J. Food Technol.* **2,** 377–397.

Pitt, J. I. (1975). Xerophilic fungi and the spoilage of foods of plant origin. *In* "Water Relations of Foods" (R. B. Duckworth, ed.), pp. 273–307. Academic Press, New York.

Pitt, J. I., and Christian J. H. B. (1968). Water relations of xerophilic fungi isolated from prunes. *Appl. Microbiol.* **16,** 1853–1858.

Pitt, J. I., and Richardson, K. C. (1973). Spoilage by preservative-resistant yeasts. *CSIRO Food Res. Q.* **33,** 80–85.

Roberts, T. A., and Ingram, M. (1966). The effect of sodium chloride, potassium nitrate and sodium nitrite on the recovery of heated bacterial spores. *J. Food Technol.* **1,** 147–163.

Scott, W. J. (1957). Water relations of food spoilage microorganisms. *Adv. Food Res.* **7,** 83–127.

Scott, W. J., and Vickery, J. R. (1939). Investigations on chilled beef. Part II. Cooling and storage in the meatworks. *Counc. Sci. Ind. Res. Aust., Bull.* **129.**

Snow, D., Crichton, M. H. G., and Wright, N. C. (1944). Mould deterioration of feeding stuffs in relation to humidity of storage. *Ann. Appl. Biol.* **31,** 102–110 and 111–116.

Stamer, J. R. (1968). Fermentation of vegetables by lactic acid bacteria. *N.Y. State Agric. Expt. Stn. (Geneva, N.Y.)* pp. 46–53.

Tonge, R. J., Baird-Parker, A. C., and Cavett, J. J. (1964). Chemical and microbiological changes during storage of vacuum packed sliced bacon. *J. Appl. Bacteriol.* **27,** 252–264.

Vaughn, R. H. (1951). The microbiology of dehydrated vegetables. *Food Res.* **16,** 429–438.

Vaughn, R. H. (1975). Lactic acid fermentation of olives with special reference to Californian conditions. *In* "Lactic Acid Bacteria in Beverages and Food" (J. G. Carr, C. V. Cutting, and G. C. Whiting, eds.), pp. 307–323. Academic Press, New York.

von Schelhorn, M. (1950). Untersuchungen über den Veberb wasserarmer Lebensmittel durch osmophile Mikroorganismen. I. Verberg von Lebensmittel durch osmophile Hefen. *Z. Lebensm.-Unters.-Forsch.* **91,** 117–124.

7

Microbial Survival

Most foods are held for short periods during processing or for long periods during storage under conditions that prevent the growth of their microflora. Often, the survival of such microorganisms can be of concern. At low temperatures, the main interest is in the survival of pathogens in frozen foods. At moderate temperatures, the survival of pathogens is relevant in foods in which growth is prevented by means other than unfavorable temperatures. The main class of food here is the dry or dried product. Finally, there is a high degree of concern for the survival of both pathogenic and spoilage microorganisms in foods being processed for commercial sterility by high-temperature treatment in hermetically sealed containers. There are also situations in food technology in which the survival of microorganisms is desirable, such as in the preservation by freezing of bacteria for use as inocula in food fermentations. Before discussing the influence of a_w on survival under these various conditions, we will consider briefly the kinetics of microbial survival.

SURVIVOR CURVES

Like its rate of multiplication, the rate of inactivation of a population of single-celled organisms tends to be exponential. Thus, the same proportion of the viable population will be inactivated in each succeeding unit of time. Consequently, a semilogarithmic plot of viable cells against time will be linear (Fig. 7.1.A), and from it a rate of "death" can be determined. This is expressed as the "decimal reduction time" or *D* value—the time taken, at the temperature and under the other conditions of the test, to reduce the number of viable cells (or, more strictly, colony-forming units) by 90%. The problem of establishing the viability of sublethally impaired cells will be referred to later.

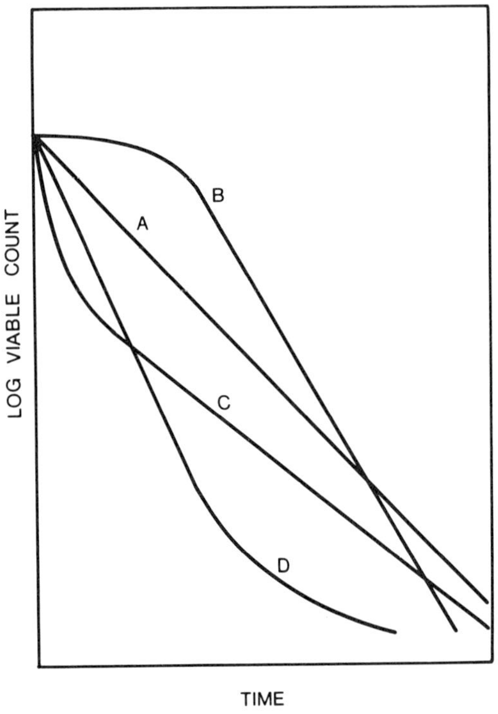

Fig. 7.1. Four types of survivor curve.

Unfortunately, semilogarithmic survivor curves are not always linear throughout. A decrease in viability may be preceded by a substantial "shoulder" in the plot due, for example, to the presence of multicellular colony-forming units, such as clumps or chains, in which every cell must be inactivated before colony-forming ability is lost (Fig. 7.1B). Such a shoulder, or even an initial increase in viable count, may result also from heat activation in a population of spores. Conversely, a rapid decrease in viability may precede a less precipitous exponential decline (Fig. 7.1C). This reflects the rapid demise of a portion of the population of greater sensitivity to the conditions due, for example, to previous injury, to a different physiological age, or to the presence of a mixed population. Finally, an exponential curve that subsequently tails off also indicates a heterogeneous population, in this case, with a small proportion possessing enhanced resistance (Fig. 7.1D).

Actual survivor curves obtained under various conditions are shown in Fig. 7.2, 7.3, and 7.4. Note that some appear to be combinations of the basic types of curve described above.

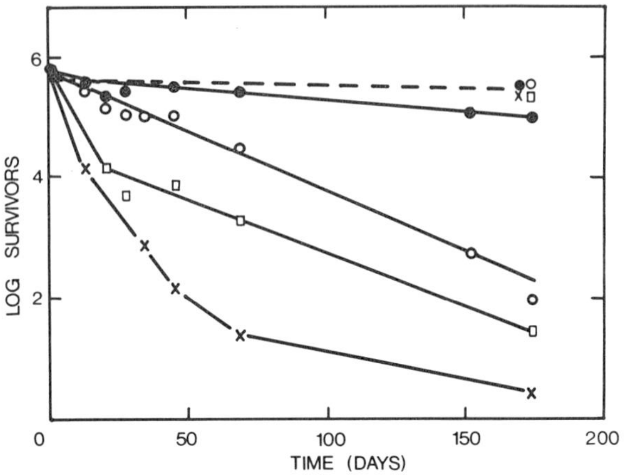

Fig. 7.2. Effect of freezing on survival of bacteria. ——, *Escherichia coli;* ---, *Bacillus mesentericus* spores; ×, −2°C; □, −5°C; ○, −10°C; ●, −20°C (Haines, 1938).

Where it is necessary, as in calculating thermal processes for canned foods, to establish an index of resistance from nonlinear curves, the time required for a substantial reduction in the numbers of viable cells of 10^6 or more to occur may be used. This has been termed the $MPED_n$ (the most probably effective heat dissipation to achieve *n overall* decimal reductions) by Mossel *et al.* (1968).

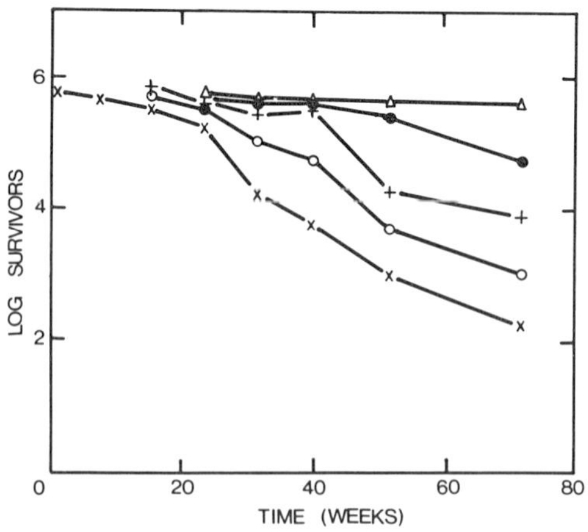

Fig. 7.3. Effect of sucrose concentration on survival of *Torula* sp. in frozen orange juice at −17.8°C. Juice containing ×, no added sucrose; ○, 10% added sucrose; +, 20% added sucrose; ●, 30% added sucrose; △, 40% added sucrose (Ingram, 1951).

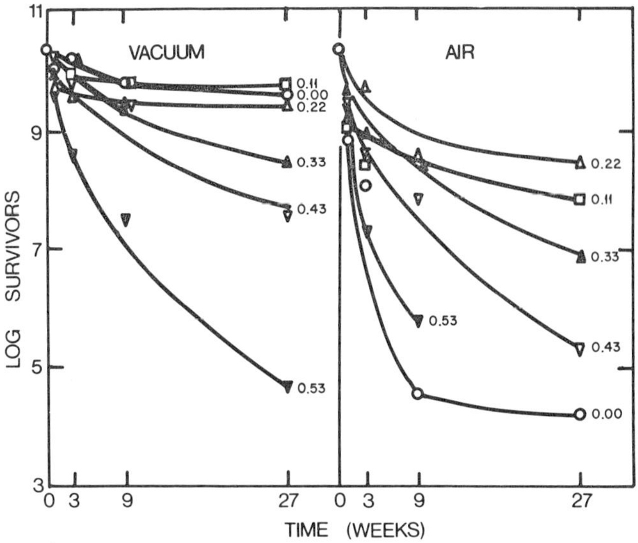

Fig. 7.4. Effect of a_w on survival, during storage at 25°C *in vacuo* and in air, of *Salmonella newport* dried in papain digest broth (Scott, 1958).

SURVIVAL AT FREEZING TEMPERATURES

Freezing and frozen storage may reduce greatly the viability of populations of sensitive microorganisms. The latter include the vegetative cells of yeasts and molds and most gram-negative bacteria. Gram-positive bacteria, especially cocci, are more resistant, and for these reasons enterococci are frequently claimed to be more suitable than *Escherichia coli* as indicators of fecal contamination in frozen foods. Many fungal spores also show this level of resistance. Least affected by freezing are bacterial spores.

The rate of freezing has been reported frequently to influence survival. A common explanation is that it influences the size of ice crystals, and, hence, the degree of mechanical damage caused to cellular structures. Although conflicting reports abound in the literature, it appears that for bacteria, if not for fungi, rapid freezing is less damaging than slow freezing.

Figure 7.2 shows the marked effect of storage temperatures, in the range −2° to −20°C, on the survival of vegetative cells. Bacterial spores retain their viability over a wide range of frozen conditions.

The recommended temperatures for storage of most frozen foods approach −20°C. At such temperatures, the chemical changes in foods are reduced to a practicable minimum. Clearly, however, they are also in a temperature range which minimizes destructive processes within microbial cells.

At temperatures in the range between the freezing point of a food and its eutectic, the a_w of the food is a function, not of its overall solute concentration, but of the ambient temperature (see Chapter 1). Thus, since temperature determines a_w, the effects of the two factors on microbial survival in the frozen state are indistinguishable. A frozen food held at −20°C has an a_w of 0.823, irrespective of its composition. However, although composition does not control a_w, it can have a marked influence on survival of microorganisms frozen in a food. Sugars, sugar alcohols, glycols, and proteins may be particularly protective, and Fig. 7.3 shows the substantial effect of added sucrose on the survival of *Torula* sp. in frozen orange juice. At −17.8°C for over 1 year, a loss of viability of about 1000-fold in the basic concentrate was reduced to barely detectable levels by addition of sucrose to a concentration of 40%.

SURVIVAL AT MODERATE TEMPERATURES

Many nonsterile foods are microbiologically stable in the moderate or ''room'' temperature range. The majority are dried or concentrated foods, owing their stability to reduced levels of a_w. Those that are rehydrated before consumption regain the ability to support microbial growth, so that the capacity of contaminating organisms to survive the period of low a_w storage is of obvious relevance.

In a study of dried milk, Higginbottom (1953) demonstrated maximal survival in the a_w range 0.05–0.20 when samples contained predominantly vegetative bacteria. Where bacterial endospores constituted the bulk of the population, survival was little affected at a_w levels down to 0.05 a_w, but was reduced substantially at 0.00 a_w.

Studies on survival of pure cultures of vegetative bacteria equilibrated to a range of a_w levels after freeze-drying have shown clearly the increase in survival that accompanies reduction in a_w level to 0.1–0.2 (Scott, 1958). Behavior at lower a_w levels (e.g., 0.0 a_w) when dried in some complex menstruums was greatly influenced by the presence of air (Fig. 7.4). Survival of *Salmonella newport* at 0.0 a_w, after freeze-drying in papain digest, was nearly maximal *in vacuo,* but was very poor when stored in air. Although *Pseudomonas fluorescens* proved more susceptible to death on storage, the qualitative response to a_w was similar.

Death of bacteria during storage at reduced a_w levels is greatly influenced by the nature of solution from which they had been dried. While nonreducing sugars are protective, reducing sugars accelerate bacterial inactivation. Scott (1958) hypothesized that reactions between carbonyl compounds and the amino acid side chains of cellular constituents were important causes of bacterial death in the dry state. He concluded that a protective medium should contain a nonreducing sugar and amino compounds that can react with any carbonyl compounds and,

TABLE 7.1.

Optimum a_w for Survival of *Staphylococcus aureus* and *Salmonella newport* in Dried Foods at 25°C[a]

Organism	Atmosphere	Cake mix	Skim milk	Onion soup	Dessert
Staphylococcus aureus	Air	0.11	0.11	0.00–0.11	0.11
	Vacuum	0.00	0.00	0.00–0.11	0.22
Salmonella newport	Air	0.11	0.11	0.00–0.11	0.00
	Vacuum	0.00	0.00	0.00–0.11	0.11

[a] Christian and Stewart, 1973.

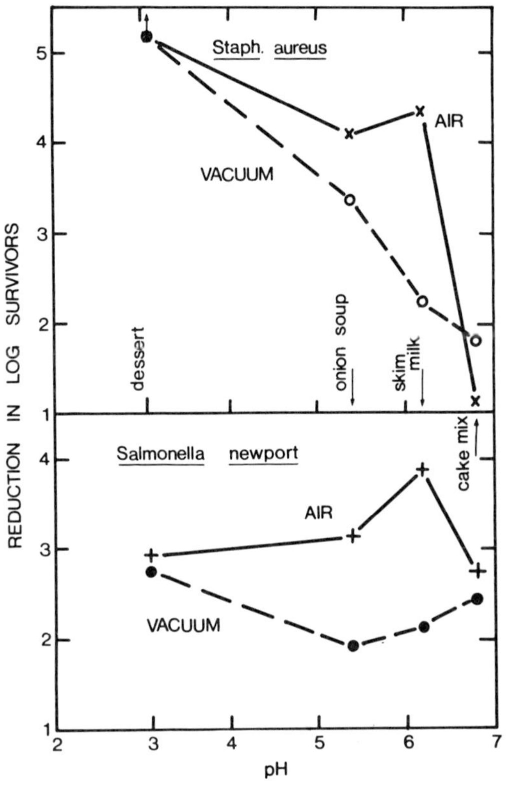

Fig. 7.5. Relationship of pH of foods to the reduction in viability of bacteria during 9 weeks storage at 25°C and 0.43 a_w in air and in vacuum (Christian and Stewart, 1973).

thus, protect cell amino groups. Marshall *et al.* (1974) subsequently suggested that a useful mixture in which to freeze-dry and store bacteria should contain 0.1 *M* sucrose, 0.2 *M* glutamate, and 0.02 *M* semicarbazide adjusted to pH 7. The recommended storage conditions were a temperature of 10°C or lower *in vacuo* at 0.1 a_w.

Similar results were obtained in a study of the survival of *Staphylococcus aureus* and *Salmonella newport* in four dried foods (Christian and Stewart, 1973). Here again, optimum survival was obtained in the range 0.0–0.22 a_w (Table 7.1). The damaging effect of oxygen was again observed, and could be duplicated with 1.5% or more oxygen in nitrogen.

Unlike salmonellae, staphylococci proved very sensitive to reductions in pH when stored in foods at 0.43 a_w (Fig. 7.5). Skim milk was more damaging to both bacteria during storage in air (but not *in vacuo*) than would be predicted from its pH value.

Survival of *Salmonella senftenberg* 775W in meat and bone meal at several temperatures between 4° and 37°C was generally greatest at 0.2 a_w, the lowest level tested (Liu *et al.*, 1969). Most rapid destruction accompanied storage at 0.9 a_w, while multiplication occurred at higher a_w levels.

Thus, while the survival of microorganisms in dried foods can be markedly affected by a_w level, there are important interactions between a_w and such factors as pH, oxygen, and food composition. For many foods, deterioration during storage in the dry state is least at the relatively low a_w levels at which microbial survival is optimum. The situation is very much like that in frozen food, where the same temperature range appears to provide optimum quality retention in foods but maximal survival by microorganisms.

SURVIVAL AT ELEVATED TEMPERATURES

Microorganisms vary greatly in heat resistance, the more resistant bacteria (e.g., *Bacillus stearothermophilus*) producing spores with decimal reduction times in neutral foods as long as 4 minutes at 121.1°C or 40 minutes at 110°C. Not all bacterial spores are this resistant, *Clostridium botulinum* type E having a $D_{110°}$ of less than 0.1 seconds (Murrell and Scott, 1966). The more sensitive, nonsporing bacteria, including pathogens, possess *D* values of about 3 minutes at a temperature of only 50°C. At high a_w and neutral pH, vegetative cells of yeasts, molds, and bacteria have *D* values that lie in the 1–15 minute-range at temperatures between 50° and 60°C.

Yeast ascospores are only slightly more heat resistant than vegetative yeast cells, but ascospores of molds can be much more so. Ascospores of *Byssochlamys* spp. are sufficiently resistant to cause spoilage problems in some acidic canned products that receive mild heat treatments. For *Byssochlamys fulva*, the $D_{88°}$ value is about 10 minutes (King *et al.*, 1969).

The qualitative effect of moisture upon microbial heat resistance is well known—moist heat is a much more effective sterilizing agent than dry heat and, wherever practicable, steam sterilization is preferred as being much more rapid than hot air (dry) sterilizing.

However, having said this, the quantitation of the effect of moisture in terms of a_w on heat resistance has proved difficult, and conflicting reports abound. Among nonsporing bacteria, salmonellae have received the greatest attention, largely because of a concern about their presence in animal feeds and in confectionery products. Their heat resistance is discussed in detail in Chapter 8. In brief, heat resistance increases as a_w decreases, the lowest a_w levels tested being, in general, about 0.1 for feeds and about 0.7 for sugars (Corry, 1971, 1974).

Water activity is also likely to be of significance in the heat treatment of foods in the intermediate moisture range. Hsieh *et al.* (1975) showed that, at pasteurization temperatures for salmonellae, staphylococci, and yeasts (50°–60°C), death rates are lowest in the a_w range 0.75–0.85 in glycerol-adjusted solutions. They recommend that any necessary heat treatment, therefore, should be given to components before combining with a_w-lowering agents.

Osmophilic yeasts respond similarly to salmonellae when heated in sucrose solutions, with decimal reduction times increasing as a_w decreases from 0.995 to 0.85 (Gibson, 1973). These organisms are more heat sensitive than salmonellae.

In some respects, reports of the water relations of the heat resistance of bacterial spores are more consistent than those of vegetative cells. Where a_w is controlled by equilibration in the vapor phase, heat resistance increases manyfold as a_w is reduced from 1.0 to a maximum in the range 0.2–0.4 a_w (Murrell and Scott, 1966; Härnulv and Snygg, 1972). Figure 7.6 shows this for *Bacillus subtilis* spores, in which the change in D value exceeds 1000-fold when a_w is controlled by vapor-phase equilibration. Adjustment of a_w level by immersing spores in glycerol solutions gave the optimal heat resistance in the same region (0.3 a_w), but with a maximum D value some 10-fold lower. Other solutes used in contact with the spores (NaCl, LiCl, glucose) had no protective effect at most a_w levels tested.

A similar study by Härnulv *et al.* (1977), using the more heat-resistant organism *Bacillus stearothermophilus,* showed that the resistance of spores was greatest in water vapor and glycerol solutions at low, but not zero, a_w. When glucose or NaCl were used to adjust a_w, only very slight variations in heat resistance were noted as a_w levels were altered. Spores inoculated in three foods; egg powder, fish protein concentrate, and wheat flour showed greater resistance at 0.33 than at 0.00 or 0.68 a_w. Solute-related effects in glycerol were related closely to a_w levels; however, in NaCl, LiCl, and glucose, solute-specific effects, distinct from those related to a_w levels, could not be discounted.

This increase in heat resistance at reduced a_w was shown by Murrell and Scott (1966) to be greatest with spores of least resistance at high a_w levels, the effect of a_w reduction being to remove most of the heat resistance differences existing

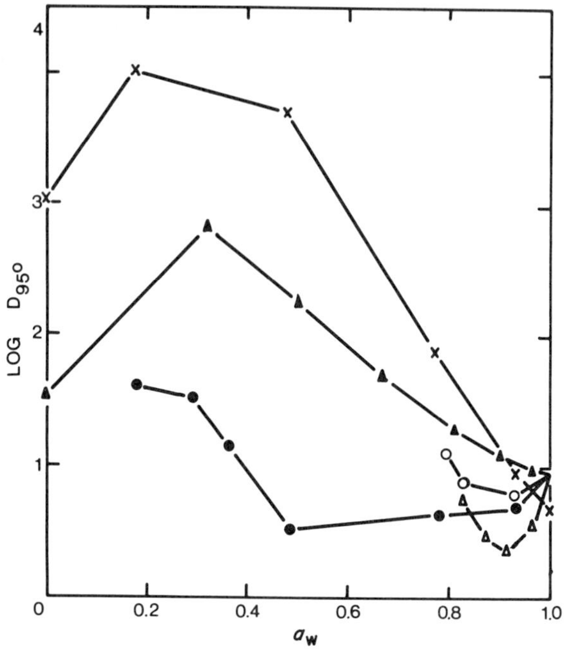

Fig. 7.6. Effect of a_w on log decimal reduction time at 95°C for spores of *Bacillus subtilis* with a_w controlled by ×, water vapor; ○, NaCl solutions; ●, LiCl solutions; △, glucose solutions; ▲, glycerol solutions (Härnulv and Snygg, 1972).

between species. Figure 7.7 demonstrates the magnitude of this effect. A difference in heat resistance of 10^5 at 1.0 a_w is reduced to 10^1 at 0.2 a_w, representing an increase in D value of 10^5 in the relatively heat-labile species (*Clostridium botulinum* type E) and of less than 2×10^1 in the heat-resistant species (*Bacillus stearothermophilus*).

Mold spores show a similar effect of a_w on heat resistance. Lubieniecki-von Schelhorn (1973) studied chlamydospores of *Humicola fuscoatra*, ascospores of *Byssochlamys fulva*, and conidiospores of *Aspergillus niger* and showed that all were more resistant to dry heat than to moist heat. Those most susceptible to high temperatures at 1.00 a_w (the conidiospores) showed the greatest gain in heat resistance as a_w was reduced toward zero. Under dry conditions, the derived $D_{100^{\circ}}$ values for the mold spores were one to two orders of magnitude lower than those reported for the most heat-resistant bacterial spores under similar conditions.

There have been many reports that vegetative microorganisms in foods containing fats or oils survive heat processes which, in low-fat systems, would be lethal. These have been reviewed by Hersom and Hulland (1963). It seems likely

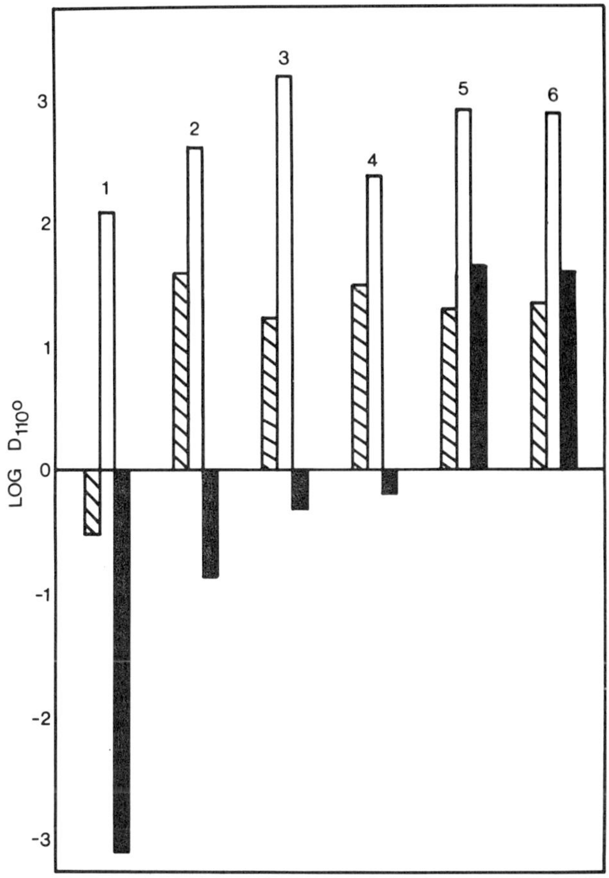

Fig. 7.7. The heat resistance of the spores of six bacteria at each of three values of a_w: (1) *Clostridium botulinum* type E; (2) *Bacillus megaterium;* (3) *Clostridium botulinum* type B; (4) *Clostridium bifermentans;* (5) *Bacillus coagulans;* (6) *Bacillus stearothermophilus;* ▧, 0.00; □, 0.2–0.4 (region of greatest resistance); ■, 0.998 (0.05 *M*-phosphate) (Murrell and Scott, 1966).

that in some situations, bacteria, frequently cocci, embedded in lipid do survive such heat processes, but their significance in such products is slight because they remain surrounded by lipid after processing and so are not able to multiply.

Similar data are available for bacterial spores. Molin and Snygg (1967) showed not only that spores of *Bacillus* spp. have greatly enhanced heat resistance in lipids, compared with those in phosphate buffer, but also that small additions of water to the oil greatly reduces this resistance. For example, the heat resistance of spores of *Bacillus cereus* in soybean oil was reduced about 1000-fold by the addition of 1% water to the oil. They point out that lipids saturated with water still have a high capacity to protect spores against heat.

The enhanced resistance to heat and the subsequent inhibition of growth are presumably both consequences of the low a_w level in the lipid. There is a need for quantitative studies of heat resistance in such systems with careful control of the ambient environmental a_w.

The basis of heat resistance of the microbial spore is not understood. Consideration of the properties and composition of the bacterial spore cortex led Lewis *et al.* (1960) to postulate that dehydration of the protoplast by contraction of the surrounding cortex confers heat resistance. More recently, Gould and Dring (1975) have suggested that the protoplast is compressed, not by contraction of the cortex, but by its expansion. Dehydration by compression appears to be accepted as the ultimate basis of heat resistance. Any theory to explain the means of this dehydration must accommodate the complicated sporulation and germination processes known to occur in these organisms.

The heat inactivation of staphylococcal enterotoxin is also influenced by a_w. Jamlang *et al.* (1971) studied the thermal destruction of enterotoxin B at NaCl concentrations of from 0.02 to 1.0 *M*. This range corresponds to an a_w range of about 0.95–0.99. The heat treatment was for 8 minutes at 70°C. At pH 6.4, the amount of toxin remaining was greater with increases in a_w. The opposite effect was observed at pH 4.5.

Troller (1976) established biphasic destruction curves for enterotoxin B heated at 149°C in Veronal buffer adjusted to 0.99 and 0.90 a_w with NaCl. Substantial inactivation occurred rapidly in the first phase, with more toxin inactivated at 0.99 than at 0.90. The second phase was one of much greater rate at 0.90 than at 0.99 a_w.

SUBLETHAL IMPAIRMENT

In situations where vegetative microbial cells are held for long periods under circumstances precluding growth or are subjected to stress from physical or chemical agents, sublethal impairment may result. This is a form of damage that prevents the cell from growing in an environment which is suboptimal, but tolerable to the normal, undamaged cell. It is often revealed as an inability to grow in the presence of selective agents or in nutritionally simple media.

One property that may be impaired is the ability to grow on media of reduced a_w. Cells of *Staphylococcus aureus* lost the ability to form colonies more rapidly on salt-containing medium than on salt-free medium during storage at 5°C (Jackson, 1974) and during heating at 60°C (Busta and Jezeski, 1965).

Sublethal impairment is a particular problem in the recovery of surviving organisms from dried and heat-processed foods, but this has been largely overcome by the use of resuscitation or preenrichment media. These usually consist of a broth based on partially hydrolyzed protein in which impaired cells may recover before transfer to a selective environment. The technique has been most commonly employed in the isolation of salmonellae from foods of low a_w.

Greater problems arise when sublethally impaired cells are to be enumerated on solid, selective media. Resuscitation in liquid medium is not recommended, as the incubation period, which will result in maximum recovery without multiplication, cannot be predicted. Two other techniques appear promising. The diluted homogenate may be plated on a nonselective agar and, after resuscitation at an appropriate temperature, is overlayed with selective agar and reincubated. Alternatively, the inoculum may be spread onto a membrane filter which is placed on a resuscitation agar medium for several hours and subsequently transferred to the selective agar for the final incubation.

The practical implications of sublethally impaired microorganisms in food have been reviewed by Busta (1975).

REFERENCES

Busta, F. E., and Jezeski, J. J. (1965). Effect of sodium chloride concentration in an agar medium on growth of heat-shocked *Staphylococcus aureus*. *Appl. Microbiol.* **11,** 404–407.

Busta, F. F. (1975). Practical implications of injured microorganisms in food. *J. Milk Food Technol.* **39,** 138–145.

Christian, J. H. B., and Stewart, B. J. (1973). Survival of *Staphylococcus aureus* and *Salmonella newport* in dried foods, as influenced by water activity and oxygen. *Microbiol. Saf. Food, Proc. Int. Symp. Food Microbiol., 8th, 1972* pp. 107–119.

Corry, J. E. L. (1971). The water relations and heat resistance of microorganisms. *B.F.M.I.R.A. Sci. Tech. Surv.* No. 73.

Corry, J. E. L. (1974). The effect of sugars and polyols on the heat resistance of salmonellae. *J. Appl. Bacteriol* **37,** 31–43.

Gibson, B. (1973). The effect of high sugar concentrations on the heat resistance of vegetative microorganisms. *J. Appl. Bacteriol.* **36,** 365–376.

Gould, G. W., and Dring, G. J. (1975). Heat resistance of bacterial endospores and concept of an expanded osmoregulatory cortex. *Nature (London)* **258,** 402–405.

Haines, R. B. (1938). The effect of freezing on bacteria. *Proc. R. Soc. London, Ser. B* **124,** 451–463.

Härnulv, B. G., and Snygg, B. G. (1972). Heat resistance of *Bacillus subtilis* spores at various water activities. *J. Appl. Bacteriol.* **35,** 615–624.

Härnulv, B. G., Johannson, M., and Snygg, B. G. (1977). Heat resistance of *Bacillus stearothermophilus* spores at different water activities. *J. Food Sci.* **42,** 91–93.

Hersom, A. C., and Hulland, E. D. (1963). "Canned Foods. An Introduction to their Microbiology," 4th ed. Churchill, London.

Higginbottom, C. (1953). The effect of storage at different relative humidities on the survival of micro-organisms in milk powder and in pure cultures dried in milk. *J. Dairy Res.* **20,** 65–75.

Hsieh, F., Acott, K., Elizondo, H., and Labuza, T. P. (1975). The effect of water activity on the heat resistance of vegetative cells in the intermediate moisture range. *Lebensm.-Wiss. u. Technol.* **8,** 78–81.

Ingram, M. (1951). The effect of cold on microorganisms. *Proc. Soc. Appl. Bacteriol.* **14,** 243–260.

Jackson, H. (1974). Loss of viability and metabolic injury of *Staphylococcus aureus* resulting from storage at 5°C. *J. Appl. Bacteriol.* **37,** 59–64.

Jamlang, E. M., Bartlett, M. L., and Snyder, H. E. (1971). Effect of pH, protein concentration and ionic strength on heat inactivation of staphylococcal enterotoxin B. *Appl. Microbiol.* **22,** 1034–1040.

King, A. D., Michener, H. D., and Ito, K. A. (1969). Control of *Byssochlamys* and related heat-resistant fungi in grape products. *Appl. Microbiol.* **18,** 166–173.

Lewis, J. C., Snell, N. S., and Burr, H. K. (1960). Water permeability of bacterial spores and the concept of a contractile cortex. *Science* **132,** 544–545.

Liu, T. S., Snoeyenbos, G. H., and Carlson, V. L. (1969). The effect of moisture and storage temperature on a *Salmonella senftenberg* 775W population in meat and bone meal. *Poult. Sci.* **48,** 1628–1633.

Lubieniecki-von Schelhorn, M. (1973). Influence of relative humidity conditions on the thermal resistance of several kinds of spores of molds. *Acta Aliment. Acad. Sci. Hung.* **2,** 163–171.

Marshall, B. J., Coote, G. G., and Scott, W. J. (1974). Some factors affecting the viability of dried bacteria during storage *in vacuo*. *Appl. Microbiol.* **27,** 648–652.

Molin, N., and Snygg, B. G. (1967). Effect of lipid materials on heat resistance of bacterial spores. *Appl. Microbiol.* **15,** 1422–1426.

Mossel, D. A. A., van Schothorst, M., and Kampelmacher, E. H. (1968). Prospects for the *Salmonella* radioiodation of some foods and feeds with particular reference to the estimation of the dose required. *In* "Elimination of Harmful Organisms from Food and Feed by Irradiation," IAEA, Vienna.

Murrell, W., and Scott, W. J. (1966). The heat resistance of bacterial spores at various water activities. *J. Gen. Microbiol.* **43,** 411–425.

Scott, W. J. (1958). The effect of residual water on the survival of dried bacteria during storage. *J. Gen. Microbiol.* **19,** 624–633.

Troller, J. A. (1976). Characterization of the thermal inactivation of staphylococcal enterotoxin B at oil-frying temperatures. *Proc. Int. Congr. Food Sci. Technol., 4th, 1974* Vol. III, pp. 189–194.

8

Food-Borne Pathogens

Perhaps the simplest and most useful definition of food-borne pathogens describes these microorganisms as those found in foods and capable of producing a pathological response in man. The list of such organisms has never been static. More sensitive methods for detecting vegetative cells, spores, and their toxic metabolites; improved epidemiological techniques; and increased awareness of food-borne diseases and the conditions that favor them have all contributed to subtractions or additions to the genera and species of bacteria considered to be pathogenic to man in foods. Geographic factors, too, have come into play. Foods that were at one time consumed only locally may now be distributed internationally and the disease-producing organisms that these foods may contain consequently achieve a much broader distribution. Thus, health authorities in a given country may be confronted frequently with food-borne diseases that they previously had encountered only rarely or not at all.

Food scientists, too, are constantly forced to reappraise their views of exactly which organisms are, or should be, considered food-borne pathogens. For example, until the discovery of turkey X disease (aflatoxicosis) in the early 1960's, the growth of molds on foods was usually thought to be harmless. The existing rule-of-thumb was simply to scrape or excise the mold-containing area and then consume the food. This practice is no longer acceptable because subsequent epidemiological studies on populations consuming relatively large amounts of moldy foods, particularly peanuts or ground nuts, have shown an increased incidence of hepatic carcinomas. The effect of a_w on growth and mycotoxigenesis by molds of this type will be covered in some detail later in this chapter.

At the same time as the definition and scope of food-borne diseases and their causal organisms are being redefined, their traditional classification into food infections and intoxications is also being questioned. Food infections are those illnesses in which viable microorganisms must be ingested, with subsequent growth of these organisms in the alimentary tract. At one time, diseases such as salmonellosis and those caused by *Clostridium perfringens* and *Bacillus cereus* were considered to produce their characteristic symptoms in this manner. Hauschild (1971) and Stark and Duncan (1972), however, have shown that *C. perfringens* does, in fact, produce an enterotoxin that is released during cell lysis in the intestinal tract. *Bacillus cereus* has also been shown (Spira and Goepfert, 1972) to produce an enterotoxin that can induce a malfunction in intestinal water–electrolyte transport in experimental animals. The status of *Salmonella* in this regard is also being questioned, as is that of food-borne *Shigella* species (Baird-Parker, 1971). Thus, the classification of food-borne pathogens as being infectious or toxigenic may no longer be valid. For this and other reasons, this chapter will consider the moisture requirements of food-borne pathogens, primarily on the basis of their growth responses to limited a_w conditions without regard to the mechanism by which illness is induced.

STAPHYLOCOCCUS AUREUS

The osmotolerance of *S. aureus* has been recognized since the initial isolation and description of the organism in 1884. Early food microbiologists, studying brining and syruping, noted that gram-positive cocci survived and grew in various curing brines, often to the exclusion of other organisms. With the subsequent discovery (Dack *et al.*, 1930) that many staphylococci are capable of producing disease symptoms following the consumption of contaminated food, impetus was added to the investigation of these organisms.

Much of the initial work on the salt tolerance of *S. aureus* was exploitative rather than exploratory. The ability of brine solutions to limit the growth of most food bacteria, while allowing proliferation of *S. aureus*, led early microbiologists to add NaCl to culture media for the isolation of this species. Hill and White (1929) described salt media containing 2–20% NaCl, which effectively inhibited a variety of gram-negative bacteria while not inhibiting gram-positive cocci. Further work on such media was conducted by Koch (1942) and Chapman (1945). The latter cited several advantages for NaCl-containing media in addition to staphylococcal selectivity: (1) improved colony pigment formation; (2) improved plasma coagulation; (3) ease of maintaining culture purity; (4) suppression of dissociation.

Chapman originally described a proteose–lactose agar containing 7.5% NaCl. However, he also performed experiments in which he added a like amount of NaCl to a gelatin medium originally proposed by Stone (1935). This medium relied on hydrolyzed gelatin ("cleared") zones surrounding colonies for the identification of staphylococci. An additional medium in which NaCl (7.5%) was added to a phenol red–mannitol agar was also tested by Chapman with good results. A subsequent modification (Chapman, 1948) of the gelatin-containing medium and the mannitol salt agar medium are still in common use.

The utility of media of this type in the testing of foods for staphylococci has been amply demonstrated; nevertheless, some suppression of staphylococci can occur in media containing 7.5% NaCl. Consequently, more recent surveys of staphylococcal isolation media (Baer *et al.*, 1966) have indicated that those not containing NaCl, such as Baird-Parker agar (Baird-Parker, 1962), are somewhat more sensitive and relate more accurately to coagulase production, a prime requisite since nearly all food-poisoning strains of *S. aureus* are coagulase positive. Baird-Parker agar also has proved to be much more efficient than salt-containing agars for the recovery of heat-damaged cells and dried cells of staphylococci (Baird-Parker and Davenport, 1965; Collins-Thompson *et al.*, 1974).

Growth

In laboratory media, the growth of staphylococci is both rapid and extensive. However, the limitation of moisture in such media by the addition of solutes profoundly limits the maximal numbers of staphylococci attained in a given medium, increases generation time, and extends the lag phase of growth. The microscopic appearance of staphylococci grown in low a_w environments is also extensively altered (Fig. 8.1). Scott (1953), working with 13 strains of *S. aureus* implicated in outbreaks of food-borne disease, found that mean growth rates are higher at a_w levels of 0.995 and 0.99 than at 0.999 a_w (Table 8.1). Below these optimal a_w levels, growth rates decrease as a function of a_w of the medium. Slow growth of 9 of the 13 strains was observed at 0.86 a_w, with total growth suppression of all strains occurring at 0.84 a_w. Reduction of a_w also caused a much more rapid decrease in growth rate under anaerobic conditions than with aerobic incubation.

TABLE 8.1.

Effect of a_w on Growth and Maximal Counts of *Staphylococcus aureus* Grown in a Laboratory Medium[a]

a_w	Growth rate (divisions/hour)	Maximal numbers cells/ml × 10^8
0.999	1.11	12.5
0.995	1.30	15.7
0.990	1.30	15.3
0.980	1.15	15.0
0.970	1.10	14.8
0.960	0.95	10.2
0.940	0.65	10.4
0.920	0.42	4.5
0.900	0.18	1.5

[a] Estimated from the data of Scott (1953).

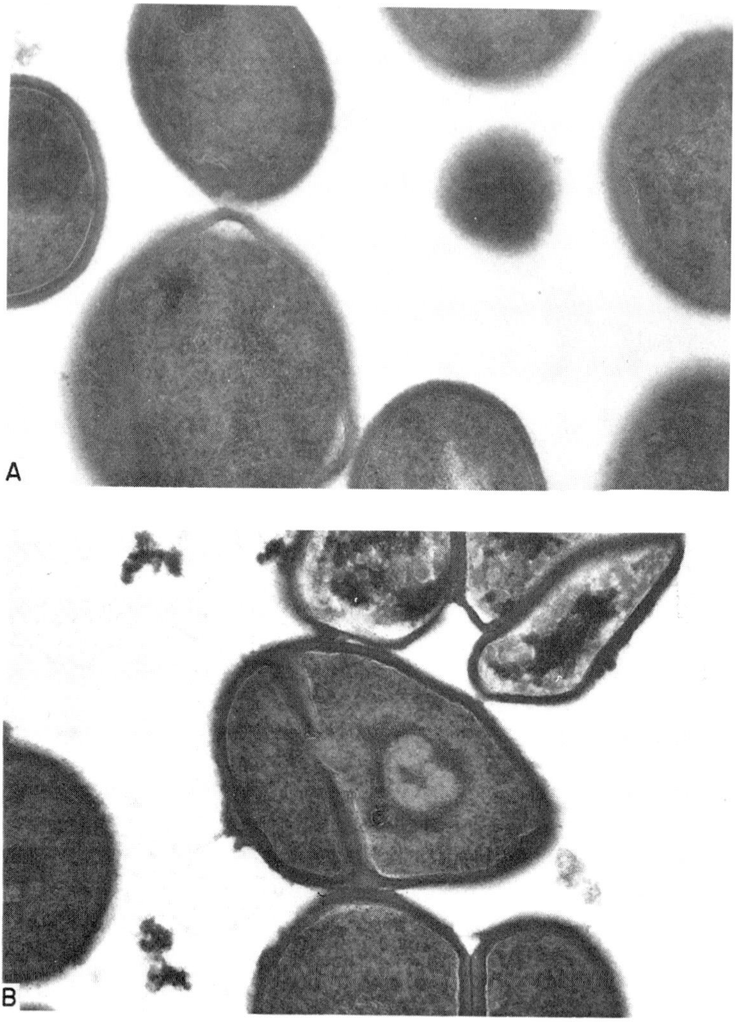

Fig. 8.1. Electron photomicrographs (× 60,000) of *Staphylococcus aureus* C243 grown in a laboratory medium at (A) 0.99 a_w and (B) 0.92 a_w.

Scott concluded that the range of a_w levels permitting growth is relatively independent of the type of solute used to attain the respective moisture conditions. These results differ, to some extent, from earlier data by Nunheimer and Fabian (1940), who found that sugars such as sucrose and glucose are somewhat more growth-limiting than NaCl at comparable (calculated) a_w levels. Differences in incubation temperature or medium composition may account for this lack of agreement. Other workers (Marshall *et al.*, 1971) have also reported instances in which various substances added to media to control moisture content produced variable effects at equivalent a_w levels, thus showing that intrinsic, solute-related effects may be substantial. For example, a much higher minimal a_w for growth was found when glycerol was the humectant (0.89) than when the a_w was adjusted with NaCl (0.86).

A typical growth curve of an enterotoxin B-producing strain of *S. aureus* in a laboratory medium adjusted to various a_w levels with glycerol is shown in Fig. 8.2. These data clearly show the inhibitory effect of a_w reduction on the growth of this strain. It should be recognized that these data pertain only to laboratory media containing relatively refined ingredients. Growth conditions found in foods may be quite different, with commensurately different effects on growth.

Generally, the growth of staphylococci may occur more rapidly in foods than in media. A corresponding increase in growth at depressed a_w levels and a marked reduction in minimal a_w levels for growth initiation may also occur. Chordash and Potter (1972) determined the growth response of a coagulase-

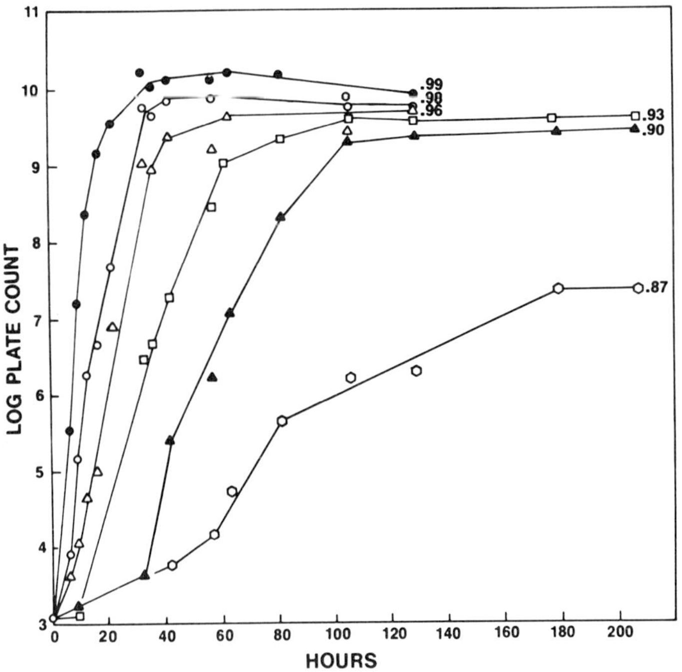

Fig. 8.2. Growth of *Staphylococcus aureus* C243 in a medium adjusted to various a_w levels with glycerol (Troller, 1971).

positive strain of *S. aureus* to a_w limitation in strained custard- and ham-containing infant foods. Higher counts were obtained for the ham product, but growth under limited moisture conditions was approximately the same in both foods. Unfortunately, only relatively high and low a_w ranges were examined in these experiments, so that the ranges at which growth response to a_w might be expected to be greatest (a_w = 0.93–0.86) were not tested.

Cured meats have been implicated frequently in outbreaks of staphylococcal food poisoning, and thus, investigations on the growth of *S. aureus* in such products and in their curing brines are more numerous than those dealing with other foods. Kelly and Dack (1936) showed that *S. aureus* can grow rapidly and selectively on salted luncheon meats containing 10% NaCl. Vigorous growth of staphylococci in inoculated ground pork and hams containing 5% NaCl plus $NaNO_3$, $NaNO_2$, and glucose also has been noted by Lechowich *et al.* (1956). Maximal counts occurring on ham surfaces were 9.5×10^8/gm, but growth in the interior of the hams, where conditions were more anaerobic, was suppressed, and counts did not exceed inoculation levels. Smoking tended to reduce surface counts, probably as a result of localized heating and drying. Staphylococci did not multiply in the curing brine and, in fact, suppression was observed; however, the addition of meat juice to these brines alleviated the suppressive effect to some extent. Anaerobic growth and enterotoxin B production were found (Genigeorgis, *et al.*, 1969) to occur in hams cured at a variety of pH, nitrite, and NaCl concentrations. Some interaction among storage temperature, salt content, and pH was noted; but even in the presence of 10^8 staphylococci per gram, hams incubated anaerobically appeared to be normal and acceptable for consumption. Like ham, bacon may support growth and enterotoxin production by staphylococci, and this product was implicated in an outbreak of food-borne disease. In this instance, an abscess containing staphylococci was found in the implicated meat, giving rise to speculation that enterotoxin was produced prior to curing and not in the cured bacon. Because this product is usually fried before consumption, it is unlikely that sufficient toxin to elicit a pathological response would remain under most circumstances.

Survival

Because staphylococci can grow at a_w levels down to at least 0.86 one might expect them to survive exposure to relatively concentrated solutions. In the case of bacteriological media (and presumably foods), some staphylococci do, in fact, survive even in the presence of 26% NaCl. However, exposure to much lower concentrations of NaCl in distilled water can be injurious without resulting in total kill of the staphylococcal population. The importance of these findings can be especially critical, not only from the standpoint of the survival of *S. aureus* in foods, but also because many laboratories rely on physiological saline (0.85% NaCl) as a diluent in the enumeration of staphylococci. The rapid decline of this organism in such diluents strongly suggests the use of a diluent containing an alternate solute such as 0.1% peptone water (King and Hurst, 1963). Distilled water solutions of NaCl (1%) frequently are used for the storage of bacteria. A storage "medium" of this type is unlikely to be satisfactory for staphylococci unless the inoculum is large enough to provide cell protection as a consequence of "carry-over" of the staphylococcal culture medium.

Maximal survival of staphylococci in a variety of dried foods was found (Christian and Stewart, 1973) to occur at 0.22 a_w. Survival decreased markedly as the a_w was increased to 0.53. This effect of a_w on the survival of *S. aureus* was modified by food constituents, oxygen tension, and pH.

Heat Resistance

The heat resistance of staphylococci at reduced a_w is an important consideration in some types of heat-pasteurized foods. Kadan *et al.* (1963) studied the effect of ingredients on the heat resistance of staphylococci introduced into several dairy products. Sucrose, when added to skim milk at concentrations up to 14%, increased thermal destruction of a pathogenic strain of *S. aureus* at 60°C. This increase in heat sensitivity was further enchanced by the addition of glucose to the sucrose-containing heating menstruum. Higher concentrations of sucrose alone, (25–57%, estimated a_w = 0.98 to 0.90) were increasingly protective to the organism. Enhanced destruction during heating (60°C) in the presence of glucose (0.950 a_w) was also observed by Calhoun and Frazier (1966), while heating in a buffer poised at 0.950 a_w with NaCl resulted in greater survival than a 0.994 a_w control. These effects can be explained by the possible formation of reducing sugar reaction products that might enhance thermal destruction. Alternatively, the a_w of a glucose solution can vary during heating, whereas the a_w of NaCl solutions is reportedly unaffected by temperature increases in the 60°C temperature range (Young, 1967); thus, the a_w of the glucose solutions during heating could have been significantly higher than that of the NaCl-adjusted menstruum.

Bean and Roberts (1975) confirmed the protective effect of NaCl in Tris-maleate buffer adjusted to 0.957 a_w (8.0% NaCl). $D_{60°}$ values increased slightly at pH 6.0, 6.5, and 7.0. A similar effect was observed at pH 6.5 in a pork macerate, also at 0.957 a_w. Hsieh *et al.* (1975) added various amounts of glycerol to skim milk and a microbiological medium (BHI), both containing an inoculum of *S. aureus* cells. Following heating at 50°, 52°, and 54°C, it was found that an increase in thermal resistance occurred with decreasing a_w from 0.99 down to the 0.76–0.85 range. This increase was reversed at < 0.76 a_w, showing a marked and steady reduction in the protective effect observed maximally at 0.76–0.85

a_w. These authors recommended prepasteurization of high a_w, intermediate-moisture food components before combination with humectants to assure maximal kill of microorganisms and so to improve microbial stability.

The effect of a_w on the thermal resistance of *S. aureus* is subject to solute-related variations. Protective effects occur predominantly with added NaCl throughout the a_w range, whereas sugars and glycerol appear to produce variable results depending upon the a_w of the cell-suspending medium. In general, it is probably wise to assume that a more extensive heat treatment will be required to produce a given level of staphylococcal destruction as the a_w of the system is lowered.

Enterotoxin Production

Staphylococci may produce one or more serologically identifiable enterotoxins in laboratory media or in foods. The fact that many salted, brined, and cured foods have, at one time or another, been implicated in outbreaks of food-borne disease clearly indicates the need for knowledge of the water relations of enterotoxin production. Before any reliable and reasonably accurate means of quantitating enterotoxins became available, food microbiologists assumed that the presence of staphylococci in high numbers in a food (10^6 or 10^7 per gram was a commonly accepted count) was sufficient evidence that the food would elicit enterotoxicosis in humans. This assumption was based on the counts of staphylococci in foods implicated in food-borne disease outbreaks. With the development of quantitative analysis techniques for enterotoxins based on serological reactions, these estimations have proven to be fairly accurate.

In most high a_w foods, more than 10^7 enterotoxigenic staphylococci per gram are sufficient to produce enough toxin to elicit a pathological response in humans. If an environmental growth factor is altered above or below optimal levels, for example temperature or pH, a corresponding decrease is noted in enterotoxin production. Often, this decrease is only commensurate with the reduction in growth of the organism, but the effect produced when NaCl concentrations in staphylococcal growth media are increased is quite different. In such circumstances, Hojvat and Jackson (1969) noted that growth and enterotoxin B production were both suppressed, the effect being greater on toxin formation than on growth. Troller (1971) related the water activity of a number of media to enterotoxin production and noted a striking decrease in enterotoxin B levels at a_w levels that produced only minimal reductions in maximal counts of staphylococci (Fig. 8.3). A similar study (Troller, 1972) with an enterotoxin A-producing strain also revealed toxin suppression, but to a somewhat lesser extent than that observed in the earlier work. This latter inhibition also appears to be relatively independent of maximal cell yield. Solute-related differences, while not altering overall suppression of toxin formation by limiting a_w levels, have also been noted

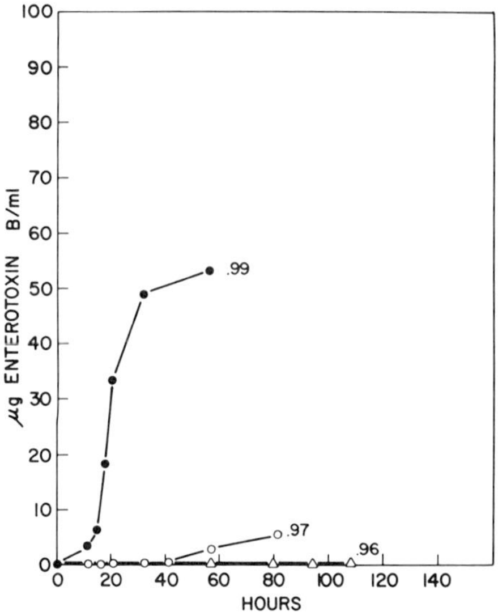

Fig. 8.3. Effect of a_w on enterotoxin B synthesis by *Staphylococcus aureus* C243 in a laboratory medium (Troller, 1971).

(Troller, 1973a). Suppression of relative or "differential" toxin production also occurs in shrimp slurries and potato doughs, suggesting that these a_w-related effects occur in foods (Troller, 1975), as well as in culture media. Further confirmation that suppression of enterotoxin B production in food occurs under conditions of low a_w can be found in the report of Niskanen and Nurmi (1976).

The mechanisms involved in the suppression of enterotoxin production at limited moisture conditions have not been investigated. The principal starting point of such studies would seem logically to be a determination of the intracellular toxin content of a_w-limited cells. Thus, the term "enterotoxin production," so frequently seen in the literature, could be defined with regard to the effect on toxin synthesis. Similarly, the effect of a_w limitation on the passage of enterotoxins through the bacterial cell membrane might prove a fruitful avenue of research. The bulk of the present evidence, much of it by inference from other species, indicates that accumulation of certain cations and amino acids (principally proline) occurs in bacteria grown at low a_w levels. Respiration rates also are depressed under these conditions, resulting in a possible depletion of energy-rich compounds essential for toxin synthesis.

At one time, enterotoxin B was regarded as a secondary metabolite; however, subsequent data have indicated that enterotoxins are produced during all growth phases (Czop and Bergdoll, 1974). Many biochemical reactions required for the survival and propagation of the cell appear to proceed with little hindrance at reduced a_w, while functions presumably less critical for survival, such as toxin formation, may be suppressed. This also may explain the suppression of the synthesis of other extracellular products such as lipase and α-hemolysin that has been observed at reduced a_w levels despite the presence of high numbers of staphylococci.

Combination Factors

Numerous factors can interact with a_w reduction to achieve an enhanced suppression of staphylococcal growth and enterotoxin production. It could be argued that this situation is the rule rather than the exception in many foods that owe their safety and stability to water limitation. Examples of such foods are ham (nitrite, pH, and NaCl) and jellies (sucrose, pH, and chemical preservatives). Some of the factors in food processing that interact with a_w limitation to produce an additive or, in some cases, an enhanced inhibitory effect on the growth of *S. aureus* are listed in Table 8.2. This table describes the general effects of the various environmental conditions found in foods and the manner in which these conditions interact with reduced a_w levels.

For further details on combination factors, the reader is referred to the reports of Christian and Stewart (1973), McLean *et al.* (1968), Genigeorgis *et al.* (1971), Nunheimer and Fabian (1940), and Heidelbauer and Middaugh (1973). Under reduced moisture conditions, the concentrations or limiting values of a given environmental factor required to produce growth and/or inhibition are narrowed. Conversely, as the concentrations or limiting levels of such factors are increased, the minimal a_w at which growth will occur is increased. Unfortunately, relatively little has been done to ascertain the effect of a_w reduction on the efficacy of chemical food preservatives, such as sorbic or propionic acid, or the interaction of other food additives, for example, antioxidants, with limited moisture conditions. The exception to this statement is a recent report by Boylan

TABLE 8.2.

Interaction of Various Factors with Reduced a_w: Effect on *Staphylococcus aureus*

Factor or process condition	Effect
Thermal process	Protection
pH	Enhanced inhibition, additive
Nitrite	Enhanced inhibition, possibly synergistic
Vacuum storage	Protection
Smoking	Principal effect probably heat (see above)

et al. (1976), who investigated the effectiveness of several antimicrobial agents in preventing *S. aureus* growth in foods in the 0.86–0.90 a_w range and at pH 5.2 and 5.6. The effectiveness of inhibitors such as the methyl ester of *p*-hydroxybenzoic acid, sodium benzoate, potassium sorbate, and calcium propionate was found to be a function of both a_w and pH. The latter is not surprising, inasmuch as salts and esters of organic acids such as these are effective only in the undissociated form, and so would be expected to be maximally effective at low pH. Specific relationships between a_w and antimicrobial effectiveness were not reported.

There is also a lack of data pertaining to the effect of reduced a_w levels on the associative growth of staphylococcal populations with other organisms in foods. Under normal circumstances (0.99 a_w), *S. aureus* is a poor competitor (Troller and Frazier, 1963); however, in limited a_w systems, it proliferates at levels that exclude other food organisms. Thus, with competitors eliminated, one would expect rapid growth of the staphylococci under conditions in which they otherwise would be suppressed. These suppositions, which seem plausible, have never been experimentally tested despite their obvious importance in maintaining the safety of many food products.

TOXIGENIC MOLDS

Toxigenic molds comprise a diverse group of fungi, the extent of which is still unknown. Many of the toxins produced by these fungi are defined only by the symptoms that they produce in domestic and/or "experimental" animals. In such cases, the potential risk that these substances pose to humans can only be inferred or, as in the case of aflatoxicosis, roughly defined by epidemiological evidence. Insofar as the authors are aware, direct experimental evidence of human toxicity of these toxins has, for obvious reasons, not been obtained.

Aflatoxin

Since the original discovery of aflatoxicosis in 1960, considerable effort has been directed at the determination of the growth factors that influence aflatoxin production. Two mold species produce aflatoxin, *Aspergillus flavus* and *Aspergillus parasiticus*. Toxin formation may occur on many substrates, but historically, epidemiologically, and economically, aflatoxicosis has assumed the greatest importance in the peanut (ground nut) industry. As a result of this emphasis, much of the pertinent research on aflatoxin formation has utilized these or related products as principal substrates.

According to Diener and Davis (1967), aflatoxin production by *A. flavus* on inoculated, sterile peanuts ceased when this product was stored at relative humidity levels < 86% (30°C). The kernel moisture level after 21 days of storage was 12–15%. In a subsequent publication, these authors (Diener and Davis, 1970) noted that *A. flavus* inoculated on Early Runner peanuts produced aflatoxin at storage (84 days) relative humidities as low as 84%. The kernel moisture content in this case was 11%; a_w levels were not reported, but were probably in the 0.83–0.84 range. Growth of *A. flavus* on laboratory media was reported to be optimal at 0.98 a_w and minimal at 0.78 a_w by Ayerst (1969).

Much of the data relating to the moisture relations of fungal growth and aflatoxin production in cereal grains refer to kernel moisture content. Because this factor is specific only for a given grain sample under specified conditions, it is not a stable parameter for experimental studies unless related to a_w by means of a sorption isotherm. Normally, peanut kernel moisture levels of approximately 15–35% are associated with maximal aflatoxin yields and minimal toxin production in the 8–12% range. Similar data pertaining to rice seem to show a minimal moisture level that is somewhat higher (18–20%) for toxin production.

Aspergilli appear to invade peanut kernels after the pods have been harvested and during curing. Because drying is usually carried out under uncontrolled climatic conditions, interruption of the drying cycle by rain and/or very high humidity can result in mold growth and subsequent aflatoxin production. If, on the other hand, drying proceeds rapidly to moisture levels at which mold growth cannot take place, fungal penetration and toxin formation are prevented. Additional care must be taken throughout peanut processing to assure that shelled, stored peanuts are not subjected to moist conditions. Condensation of moisture in storage vaults or on the surface of bags containing shelled peanuts following their removal from cold storage may create conditions in which the peanuts are particularly susceptible to mold growth, with consequent production of aflatoxin (Fig. 8.4).

The effect of NaCl or other salts on growth and aflatoxin formation by aspergilli in laboratory media has been investigated by Shih and Marth (1972), who found toxin production to be enhanced by the addition of 1–2% NaCl (estimated a_w = 0.99). The increase in toxin concentrations was attributed to increased synthesis and to enhanced release of toxin into the medium. Further increases in NaCl concentrations suppressed this transfer but stimulated toxin synthesis, the net result being progressively lower concentrations of aflatoxin in media supplemented with 3–5% NaCl (estimated a_w 0.98 to 0.97). Depending on the mold species tested and the temperatures of incubation, further increases in NaCl concentrations were inhibitory to growth and toxin production, with total inhibition of toxin synthesis observed at 14% NaCl (estimated a_w = 0.90).

Many preserved or semipreserved foods such as cheese or sausages may contain 1–2% NaCl (although NaCl levels in the water phase of such products may be much higher) and thus be extremely susceptible to aflatoxin contamination. In fact, experimentally inoculated products of this type rarely support appreciable

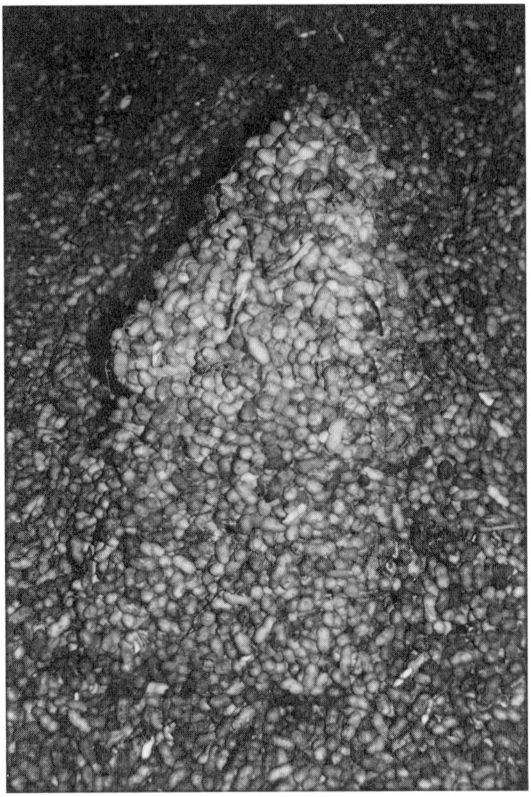

Fig. 8.4. Mold-agglomerated peanuts. These "soldiers" occur as a result of poor ventilation in storage structures with resulting condensation on ceiling joists. The condensate drips into the stored peanuts forming localized areas of high moisture which support mold growth (photo courtesy of M. Dickson).

production of aflatoxin. This apparent paradox may be attributed to the absence of suitable nutrients, the presence of solutes that might reduce the solubility of O_2 in the food, the presence of inhibitory compounds generated in the product (such as fatty acids), or inhibition by other organisms normally inhabiting these foods, such as lactic acid-producing streptococci and bacilli. Northolt *et al.* (1976) noted that the growth of *A. parasiticus* was more rapid at a given a_w if glycerol, rather than a mixture of salts, was used to adjust the a_w of the medium. In the glycerol medium, growth was observed at 0.82 a_w, but not at 0.79, and toxin production occurred at 0.87 a_w, but not at 0.82.

The heat (55°C) resistance of conidiospores harvested from toxigenic strains of *A. flavus* and *A. parasiticus* was studied in a phosphate buffer by Doyle and

Marth (1975). The extent of conidial sensitivity to thermal stress in the presence of NaCl, glucose, or sucrose was solute dependent and was influenced strongly by the solute concentration or a_w level of the heating menstruum. A solute-related protective effect was observed as the a_w was lowered; however, at 0.94 a_w, the protection exhibited by NaCl exceeded that of sucrose or glucose. These authors suggested that the type of solute used is equally as important as the a_w level in determining thermal resistance.

In summary, the minimal a_w for growth of aflatoxigenic molds is probably at or near 0.80 a_w. Toxin production probably ceases at 0.83 a_w.

Ochratoxin

Ochratoxins A, B, and C are a group of chemically related dihydrocoumarin derivatives produced by a number of species of the genera *Aspergillus* and *Penicillium*. All three ochratoxin types are synthesized on a variety of laboratory media; however, only ochratoxin A and B occur naturally. Ochratoxins are synthesized on a variety of cereal substrates and owe their toxicity to the produc-

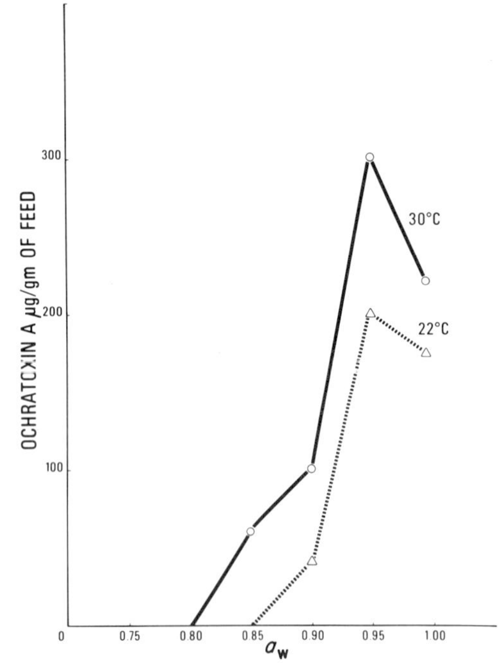

Fig. 8.5. Effects of a_w and temperature on the production of ochratoxin A by *Aspergillus ochraceus* on poultry feed (Bacon *et al.*, 1973).

tion of fatty deposits in the liver and kidneys of experimental animals. No direct evidence of ochratoxicosis in humans exists.

A strain of *Aspergillus ochraceus* (Van Walbeek *et al.*, 1968) produced ochratoxin A in a medium containing 15% sucrose (estimated $a_w = 0.99$). Bacon *et al.* (1973) have reported the minimal a_w for production of this toxin in poultry feed to be 0.85 at 30°C, with greatest production occurring at 0.95 a_w. Water activity levels greater than 0.95 tended to suppress ochratoxin A production (Fig. 8.5). An a_w of 0.90–0.93 was found by Harwig and Chen (1974) to be the most favorable for ochratoxin A production by *Penicillium viridicatum* on wheat and barley.

Penicillic Acid

Like ochratoxin, this mycotoxin is produced by species of the genera *Penicillium* and *Aspergillus*. Penicillic acid is a carcinogenic lactone, which may cause acute symptoms, such as coma and convulsions, in mice and guinea pigs. Production in poultry feed occurs at 0.80 a_w and above, with greatest yield at 0.90 a_w

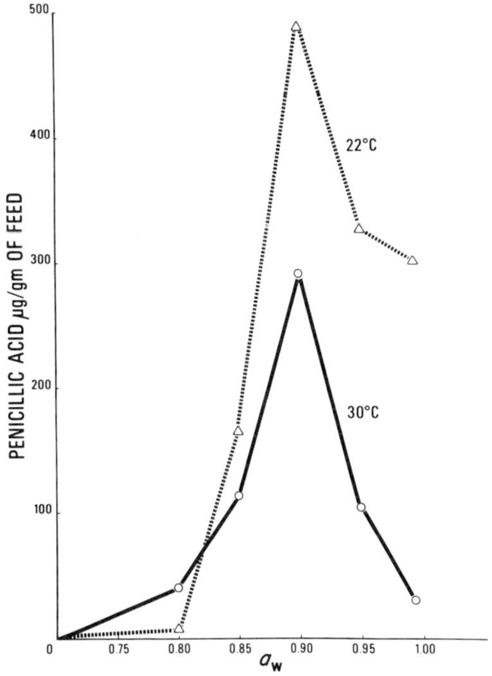

Fig. 8.6. Effects of a_w and temperature on the production of penicillic acid by *Aspergillus ochraceus* on poultry feed (Bacon *et al.*, 1973).

(90% moisture). Incubation at a_w levels > 0.90 tends to yield less toxin (Fig. 8.6). Water activity levels for maximal production are strongly temperature-dependent (Bacon *et al.*, 1973).

Patulin

This toxic antibiotic is produced by several species of *Aspergillus* and *Penicillium* and by *Byssochlamys nivea*. It is believed to possess both carcinogenic and teratogenic properties, as well as exhibiting acute toxicity in experimental animals. The organisms synthesizing patulin are common storage rot molds of pomaceous fruits. Patulin also has been detected in cattle feeds. To date, apple sap, apple juice, and apple cider have been found to be the only human foods naturally contaminated with this mycotoxin.

Optimal a_w for toxin production has not been reported. Optimal and minimal a_w levels for growth of some (but not all) patulin-producing molds are shown in Table 8.3. The ability of these species to grow and produce toxin in rotten fruit tissues (about 0.995 a_w) would, however, indicate that very high a_w levels do not greatly inhibit patulin production. Furthermore, detection of the toxin in heated apple juice would indicate that the toxin remains, to at least some degree, stable when heated in high a_w, acidic environments.

Experimentally inoculated foods have been analyzed for the presence of patulin. Fermented, dried sausage inoculated with *Penicillium expansum* (probable a_w = 0.90 to 0.94) did not support the production of this mycotoxin, although it was detected in experimental lots of cheddar cheese (probable a_w = 0.91 to 0.93) stored at 25°C for 2 weeks (Stott and Bullerman, 1975). Germination of *P. expansum* spores has been reported to occur at 0.84 a_w in a laboratory medium (Orth, 1976).

TABLE 8.3.

Water Activity Minima and Maxima for Growth of Some Patulin-Producing Molds[a]

	a_w	
Culture	Minimum	Optimum
Aspergillus fumigatus	0.82	0.97
Penicillium cyclopium	0.82	0.98
Penicillium expansum	0.83	—
Penicillium lanosum	0.82	—
Penicillium patulum	0.81	—
Penicillium urticae	0.84	—

[a] Ayerst, 1969; Orth, 1976.

Stachybotryn

This toxin, synthesized by *Stachybotrys atra*, has been reported to produce necrosis of mucous membranes in horses. The limiting and optimal a_w levels for production of stachybotryn in a laboratory medium are 0.94 and 0.98, respectively (Jarvis, 1971).

Sterigmatocystin

Only two reports appear to exist on the effect of a_w on formation of sterigmatocystin, a toxic (experimental animals) metabolite of *Aspergillus versicolor*. Both of these relate to meat or meatlike products. Halls and Ayres (1973) reported the production of this mycotoxin in inoculated experimental, country-cured hams, which had a_w values estimated to be in the 0.95–0.97 range.

Mixed and individual strains of *A. versicolor* were grown on laboratory media and minced pork adjusted to 0.99 to 0.90 a_w by Incze and Frank (1976). Although growth of this organism occurred at all a_w levels tested, sterigmatocystin was not found under any conditions.

The minimal a_w for germination of conidiospores of *A. versicolor* was determined by Orth (1976) to be 0.84. The next lowest a_w tested, 0.77 a_w, did not support germination in Czapek–Dox agar incubated at 25°C.

As stated above, mycotoxigenic molds represent a risk of unknown dimensions to humans, and, for this reason, their growth should be prevented or curtailed in our food supplies. Various environmental factors influence growth and toxin production, but moisture limitation is probably the control measure most frequently employed. These molds represent a remarkably diverse group of organisms with equally varied nutritional and environmental requirements. However, some general conclusions can be drawn with regard to moisture conditions

TABLE 8.4.

Optimal and Minimal a_w Limits for Growth of Some Mycotoxigenic Fungi[a]

Culture	a_w Optimum	a_w Minimum
Aspergillus chevalieri	0.93	0.71
Aspergillus flavus	0.98	0.78
Aspergillus fumigatus	0.97	0.82
Aspergillus candidus	0.98	0.75
Penicillium martensii	0.98	0.79
Penicillium islandicum	0.79	0.83
Stachybotrys atra	0.98	0.94

[a] From Ayerst (1969) and Jarvis (1971).

necessary to limit their growth. Generally, a food in the 0.65–0.70 a_w range (Table 8.4 and Appendix B) can be considered refractory to mold growth and mycotoxin formation. Cereal grains containing 8–10% moisture will normally fall in this range. The practicalities of many food processes may sometimes rule out the attainment of these moisture levels, but it should be recognized that the amount of available moisture, in the absence of other controlling factors such as pH and temperature extremes, can almost certainly be related to the amount of microbial growth in a given food system. At present, only fragmentary information exists on the effect of a_w on mycotoxin formation and stability in human foods.

SALMONELLA

The genus *Salmonella* consists of more than 1500 serologically defined species or serotypes; however, relatively few of these have been implicated in food-borne disease outbreaks in humans. The foods most frequently incriminated in outbreaks of salmonellosis include red meats (usually inadequately cooked) and poultry products. This pattern appears to exist throughout the world. The actual incidence of this disease is very difficult to estimate, principally because of rather haphazard reporting of suspected outbreaks to public health agencies. Some experts have estimated that from 90 to 99% of the actual number of cases in the United States alone (Communicable Disease Center, 1974) are unreported. The symptoms of this disease appear within 24–48 hours following ingestion of contaminated food, and consist of vomiting, abdominal pain, diarrhea, and often fever. These symptoms seldom linger beyond 5 days, and recovery is commonly uneventful; however, in the case of the aged or very young (< 4 years of age), the mortality rate from this disease can be significant.

Growth

Numerous studies of the growth of salmonellae in low a_w laboratory media and foods exist in the literature. Much of this emphasis, of course, stems from the need to both establish optimal growth conditions for recovery of salmonellae, as well as to develop procedures that might prevent growth and contamination of foods.

The minimal a_w for growth of three *Salmonella* species was found by Christian and Scott (1953) to be 0.941, although one of the three test strains appeared to be inhibited at 0.945 a_w. These data were obtained using laboratory media containing either salts or sucrose to control a_w. Minimal growth levels were found to be slightly lower (0.93 a_w) in three foods tested by these authors. Nutritional depletion of the growth medium by the use of a defined, glucose–salts medium, raised

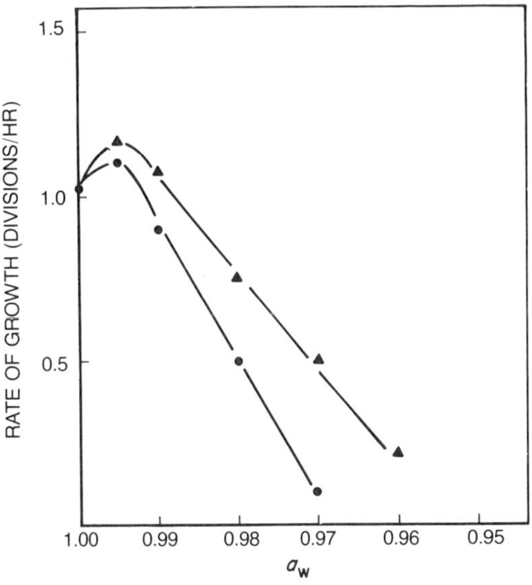

Fig. 8.7. Effect of a_w on growth rate of *Salmonella oranienburg* using a mixture of NaCl, KCl, and Na_2SO_4 (●) and glycerol (▲) to control a_w (Christian, 1955).

the minimal a_w for growth to 0.97 (Christian, 1955). Supplementation of such media with mixtures of vitamins and amino acids extended the minimal a_w for growth to the 0.95–0.94 range. Generally, the lower range of a_w levels permitting growth was independent of the solutes used for a_w adjustment. Other workers (Clayson and Blood, 1957), using surface equilibration techniques, found slightly lower a_w levels (0.92) as minimal for *Salmonella* growth.

Growth determinations of 15 *Salmonella* strains revealed (Christian and Scott, 1953) that maximal rates can be obtained at 0.995 a_w, with significantly lower growth rates at 0.999 a_w and 0.990 a_w. Rates decreased steeply at lower a_w levels, whether salts, sugar, or miscellaneous nutrients were the predominant solutes controlling a_w, although some slight solute-related effects were noted (Fig. 8.7). Similarly, decreases in a_w below 0.99 extended lag times in laboratory media. Growth rates also were suppressed in three of the test strains when incubated at various a_w levels under anaerobic conditions.

Samples of dried pork and beef, rehydrated to various moisture levels, did not support the growth of an inoculated *Salmonella* strain at 10, 20, 30, 40, and 50% moisture. However, growth did occur in two of the beef samples adjusted to 60% moisture (Segalove and Dack, 1951). Other workers (Blanche Koelensmid and van Rhee, 1964) have noted that the maximal NaCl concentration permitting growth of salmonellae in a medium and in ham "jelly" lies between 7 and 8% (corresponding to about 0.95 a_w). Survival of this organism at high (20%) NaCl concentrations was enhanced in the ham jelly, possibly as a result of a protective effect of the high protein content.

Probably more is known about the physiological circumstances occurring as a result of growth of salmonellae at low a_w than of any other organism or group of organisms. Two of the principal metabolic characteristics exhibited by salmonellae growing under reduced moisture conditions are the specific accumulations of K^+ and of proline. In cells respiring in the presence of glucose and proline, the intracellular potassium pool formed is larger than the amino acid pool at a_w levels down to 0.97. At lower a_w levels, the amino acid pool becomes dominant. Christian and Hall (1972) suggest that the osmotically active pool formed by the rapid uptake of proline at low a_w levels reverses the plasmolysis and membrane shrinkage that occur in concentrated environments. The restoration of the cell membrane to the configuration that it assumes at higher a_w levels could account for the increased respiration rates (Christian and Waltho, 1966) occurring in proline-supplemented, low a_w cultures of *Salmonella oranienburg*.

Virtually all of the reported data on the effect of moisture conditions on the survival of salmonellae have pertained to animal feeds containing bone and fish meals and, to a much lesser extent, cereal products. This emphasis stems largely from the current interest in eliminating these organisms from animal feeds in an attempt to break the *Salmonella* "cycle" of infection in domestic animals. Bone meals adjusted to moisture levels of 5–30% exhibited increasing death of *Salmonella* populations at 30°C as the amount of water was increased. However, these populations remained relatively stable at 5°C under similar moisture conditions. A 5-log reduction in viability was observed in 15% moisture meal (a_w = 0.82) held for 72 hours at 50°C (Liu *et al.*, 1969). Other animal feeds were tested for their ability to protect inoculated salmonellae by Mossel and Koopman (1965), who concluded that the primary lethal effect in fish and meat meals and in casein (a_w = 0.46) is caused by osmotic shock, although substrate composition and type of inoculum are also factors.

Heat Resistance

These and other findings that relate the survival of salmonellae to the moisture conditions of the substrate have been extended to the effect of a_w on the heat resistance of these organisms. In laboratory media adjusted to various a_w levels with sucrose, Gibson (1973) found that the heat resistance of *Salmonella senftenberg* 775 W gradually increases as the a_w of the suspending menstruum decreases (Fig. 8.8), whereas similar experiments with *Salmonella typhimurium* indicated an irregular trend toward decreasing heat resistance. The observed increase in heat resistance with the *Salmonella senftenberg* strain was found (by observation of electron photomicrographs) to be accompanied by marked cell volume reductions as the a_w of the cell suspensions was decreased. It was specu-

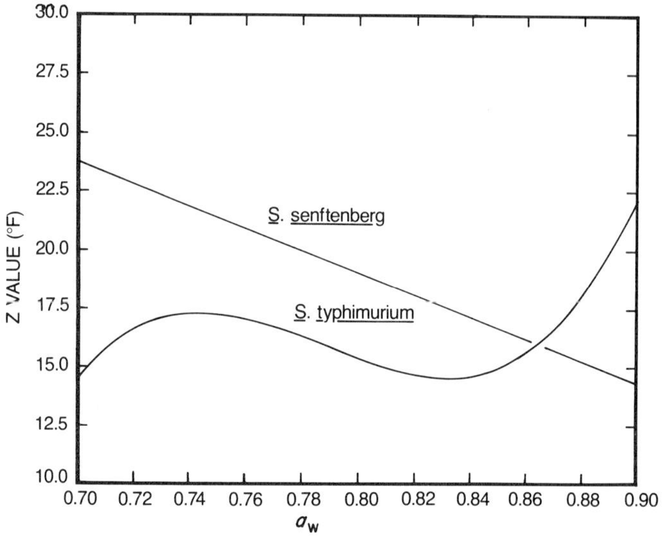

Fig. 8.8. Effect of a_w on the thermal stability of *Salmonella typhimurium* and *Salmonella senftenberg*. Data replotted from Gibson (1973) as a polynomial regression.

lated that the water content of the cell was commensurately reduced (see also Christian and Hall, 1972; Corry, 1974), thereby conferring a greater degree of heat protection to cellular proteins. These data have been questioned by Corry (1974), who found much higher $D_{65.5°}$ values than Gibson. These differences were attributed to the heating method used by Gibson, who added suspensions of salmonellae directly to hot sugar solutions, thereby exposing the organisms to thermal and osmotic shock simultaneously.

The extent to which reduced moisture conditions protect salmonellae is of special concern to the confectionery industry. The manufacture of candies may require the use of ingredients that carry a relatively high *Salmonella* risk, such as nonfat dried milk, certain dyes, egg products, and cocoa. Many candy manufacturing processes involve only a nominally lethal heat treatment. The extent to which salmonellae could be protected by the high sugar content of such products must be considered. Goepfert *et al.* (1970) have noted that not only the amount of solute but also the type of solute present is important in determining the heat stability of salmonellae. Sucrose is clearly more protective than several other sugars (Table 8.5), and glycerol is the least protective. It is interesting to note that when invert sugar was compared to sucrose in heat studies conducted at identical a_w levels (0.96), the D value of three *Salmonella* strains was lowered dramatically. Similar, although smaller, D value reductions were observed when 1% glucose was added to a 40% sucrose solution (Foster *et al.*, 1970). Thus, it may be possible to reduce public health risks associated with *Salmonella* con-

TABLE 8.5.

Effect of Various Solutes on the Heat Resistance of *Salmonella montevideo* at a_w = 0.96, 57.2°C, pH 6.9[a]

Controlling substance	$D_{57°}$ (minutes)
Sucrose	16.5
Sucrose + glucose (40:1)	9.0
Glycerol	1.2
Fructose	1.3
Sorbitol	5.5

[a] From Goepfert *et al.* (1970) and Foster *et al.* (1970).

tamination of confectionery products by the simple expedient of utilizing mixed sugars (This is already the case for many types of candies).

The heat resistance conferred on *Salmonella* cultures by limited moisture conditions has also been described by Garibaldi (1968) for egg yolk, by McDonough and Hargrove (1968) for dehydrated milk, and by Riemann (1968) for animal feeds. Baird-Parker *et al.* (1970), however, state that in artificial media, the heat resistance of salmonellae is not directly relatable to a_w, but rather, depends more on the solute employed. When sucrose was used, all strains showed increasing heat resistance with increasing solute concentrations, but differences were seen with solutes such as NaCl and glycerol, within the a_w range 0.85–0.98, depending upon whether or not the test strains were heat resistant. These authors concluded that the effects of limited moisture conditions on the thermal stability of salmonellae are difficult to predict or extrapolate from medium to medium or from food to food.

Corry (1974) examined the effect of various polyols and sugars on the heat resistance of three strains of *Salmonella* and noted that a nonlinear relationship exists between $D_{65°}$ value and a_w. A linear relationship was found, however, between $D_{65°}$ and solute concentration (percent w/w), except when glycerol was employed as the a_w-controlling solute. $D_{65°}$ values were always much lower with glycerol than with sugars. Electron photomicrographs of glycerol-treated *Salmonella* cells showed that they underwent little or no plasmolysis compared to sugar-treated cells (especially those exposed to sucrose). This author speculated that glycerol readily enters the cell to form an intracellular glycerol–water solution which, by some mechanism, is more protective than water alone, but is not as protective as when water is removed, as in the case of sucrose-plasmolyzed cells. In these studies, the extent of plasmolysis was measured by electron photomicrographs; however, in later work, Corry (1976) used a more precise, turbidometric method to verify the relationship between heat resistance and degree of cell penetration by solutes (lack of plasmolysis). These data show a close correlation between turbidity and the extent of protection conferred by sucrose, glucose, fructose, sorbitol, and glycerol. Although solute concentrations were quite low ($\leqslant$ 0.5 molal; a_w = > 0.99), this author concluded that heat resistance is more a function of cell dehydration than the replacement of intracellular water by solutes. In any case, a_w does not at present appear to be integrally involved in these effects.

Obviously, the relationship between moisture conditions existing in a given food and the amount of heat required to destroy salmonellae is complex. Such factors as the types of solutes present, species of *Salmonella*, type of food ingredients, solute concentration, and a_w may assume important roles in determining the thermolability of these organisms.

CLOSTRIDIUM PERFRINGENS

Clostridium perfringens, or *Clostridium welchii*, as it is sometimes called, is a normal inhabitant of the human intestinal tract. In addition, this organism is routinely cultured from the carcasses of beef, swine, and poultry, and it is probably by this means that *C. perfringens* enters the food chain. Although recognized as a food-borne pathogen in European countries for many years, it is only recently that the universal occurrence of this disease has been recognized. Broader recognition of perfringens food poisoning has resulted in steadily increasing incidence rates. However, it is estimated that less than 5% of the actual number of outbreaks are reported. In addition, changing food consumption patterns, such as the increased popularity of foods prepared by food service organizations, also have had a significant effect on the occurrence of this disease.

Spencer (1969) has described *C. welchii* food poisoning as "a pathological condition of man resulting from the ingestion of food in which large numbers of certain types of *C. welchii* have grown." Clinically, symptoms of this disease appear within 6–18 hours following ingestion of contaminated food and most commonly consist of abdominal pain and cramps with diarrhea. These symptoms persist for 24–36 hours without pathological sequelae. The foods most frequently implicated in outbreaks of this disease are cooked meat and poultry products that have been maintained at temperatures conducive to growth of *C. perfringens* following cooking or processing. Because the organism grows over a temperature range of 12°–52°C, practical control is achieved by prompt cooling of cooked foods and the maintenance of adequate refrigeration.

Clostridium perfringens food poisoning results from an infectious process requiring high numbers of the causative organism (10^7/gm are commonly thought to be required) to produce the disease symptoms. Reports by Stark and Duncan (1972) and Hauschild (1971) have clearly established that these organisms form one or more proteinaceous, heat-labile toxins capable of producing disease symp-

TABLE 8.6.

Effect of a_w on Generation Times of *Clostridium perfringens* NCTC-8797[a]

a_w	Generation time (minutes)		
	Glycerol	Sucrose	NaCl
0.995	22	43	43
0.98	40	51	43
0.97	42	67	63
0.95	62	N.G.	N.G.
0.93	N.G.[b]	N.G.	N.G.
0.91	N.G.	N.G.	N.G.

[a] Data are estimated from the published figures of Kang *et al.* (1969).
[b] N.G., no growth within 72 hours at 37°C.

toms in laboratory animals and presumably in humans. Toxin synthesis appears to be related to sporulation.

Clostridium perfringens is one of the most rapidly growing microorganisms known. Generation times in the 10- to 11-minute range are frequently recorded under optimal conditions. Unfortunately, none of the existing reports details the effect of a_w limitation on growth rates; however, generation times calculated from the data of Kang *et al.* (1969), who used glycerol, sucrose, and NaCl as solutes to adjust a_w levels, clearly show a reduction in growth rates as the a_w is lowered (Table 8.6). At equivalent a_w levels, glycerol appears somewhat less inhibiting than either sucrose or NaCl. These results agree with those of Kim (1965) who, using similar solutes, determined that the limiting a_w for growth of *C. perfringens* is between 0.97 and 0.95 when sucrose or NaCl is employed as solute and between 0.95 and 0.93 with glycerol. The minimal a_w permitting sporulation (0.97 or above) was somewhat higher than the a_w levels permitting growth. Strong *et al.* (1970) also investigated the effect of various solutes on the growth of *C. perfringens* and noted that the lowest a_w allowing growth is obtained with glucose (0.96) and that KCl and NaCl are somewhat more inhibitory. The longest lag times were noted with KCl, but as a_w was lowered, both the rate and amount of growth are reduced. Other studies in which NaCl was added to media (Oxhøj, 1943; Gough and Alford, 1965) indicate that 6% NaCl is required to prevent the growth of *C. perfringens*. However, this organism appears to survive smoking in hams pumped with a 17.9% solution of NaCl plus nitrate and nitrite. Although spore germination and outgrowth probably would not occur under these conditions, it is obvious that processors of cured meats cannot rely on reduced moisture conditions per se to assure the absence of *C. perfringens* from their product at palatable NaCl concentrations.

A number of factors may interact with reduced a_w levels to enhance inhibition of this organism. For example, a_w adjustment to the extremes for growth of *C. perfringens* is generally thought to be much more inhibitory when the pH of the medium approaches the minimal level for growth. There is, however, a report (Strong *et al.*, 1970) that indicates quite the opposite, namely, that at low a_w levels, the inhibitory effect of low pH appears to diminish. These authors attribute their findings to more than one pH optimum for growth or perhaps to a very broad optimal pH range. Another "combination" factor that should be considered is the influence of the oxidation–reduction potential (Eh) of the medium on the moisture requirements of *C. perfringens*. This organism is not considered an obligate anaerobe, since it is capable of initiating growth in the +200 to +250 mV ORP range under optimal conditions. If, however, the NaCl content of the medium is increased to 5% (Mead, 1969), growth will occur only if the medium is poised at much lower Eh values (+97 mV). A similar effect on the survival of this organism was noted. These results could be important in the selection of packaging materials for foods that are susceptible to the growth of *C. perfringens*. In such cases, oxygen permeability of a packaging film plus a low (about 5%) level of NaCl could give a degree of protection that neither factor could provide singly. Insofar as these authors are aware, the data of Mead (1969) provide the only indication of such a possibility, a finding which appears to be worthy of further investigation. Research on other food-borne pathogens, especially *C. botulinum,* would also be important in this context.

CLOSTRIDIUM BOTULINUM

Since its first isolation from a raw, salted ham in 1896, *C. botulinum* has been associated with cured meats and, in fact, its name is derived from *botulus,* the Latin term for sausage. For a time, botulism, or Kerner's disease as it was once known, was thought to be associated solely with sausages of various types. However, by the end of the nineteenth century, other types of foods, including canned beans and mayonnaise, had been incriminated in botulism outbreaks in Europe and North America.

Clostridium botulinum is an obligately anaerobic bacillus that forms heat-resistant spores. Six serologically identifiable exotoxins are formed by strains of this organism. These toxins, identified by letters of the alphabet, are highly potent proteins. Thus far, toxins labeled A, B, C, D, E, and F have been recognized; however, C and D have no known toxicity for man and usually produce disease in birds (primarily ducks and gulls) or in cattle. In North

America, botulism is reputed to be caused most commonly by type A toxin, whereas in European countries type B toxin (produced by nonproteolytic strains of *C. botulinum*) is more frequently incriminated. In addition, geographic areas of the world with a high consumption level of fish and fish products (Japan, Scandinavia, arctic areas) are apt to have higher incidence rates of type E botulism.

The growth of *C. botulinum* in foods may not manifest itself in obvious, deleterious, organoleptic changes, and for this reason the rejection of off-flavor or off-odor foods will not ensure safety. The symptoms of this disease are quite dramatic, and include blurred vision, mydriasis, and difficulty in swallowing. Death usually occurs as a result of respiratory failure. Gastrointestinal symptoms such as vomiting normally precede neurological symptoms and occur within 10–36 hours after ingestion of a toxin-containing food. Symptoms persist for 5 or 6 days before death occurs, and the mortality rate is very high.

Although relatively rare, the extremely high mortality rate, the ubiquity of the causative organism, and the insidious nature of this disease have caused it to claim the attention of numerous investigators since its discovery. A number of these studies have dealt with the effect of limited moisture environments on growth, sporulation, spore germination, and toxin production in laboratory media and in food systems.

Growth

Despite the early association of botulism with cured meats, this disease has been associated with home and commercially canned foods (U.S.A.), smoked fish (U.S.A. and U.S.S.R.), fish and rice (Japan), and a variety of other products. Products of these types do not normally rely on reduced a_w to prevent the growth of *C. botulinum*. However, salted foods have been implicated in outbreaks of botulism in many parts of the world, e.g., salted fish in Scandinavian countries and the U.S.S.R., beef jerky in the United States and Northern Canada, and hams and other cured meats in Europe.

Looking to the future, the present concern at the possible conversion of nitrites to potentially carcinogenic nitrosamines may lead to restrictions on nitrite addition to cured meats. Since nitrite appears to play a prominent role in suppressing the outgrowth of spores of *C. botulinum,* its prohibition would place greater emphasis on the use of limited moisture conditions to preserve these foods. For these reasons, studies on the growth of *C. botulinum* in low a_w environments assume considerable practical importance.

Inhibition of the growth of *C. botulinum* normally occurs at NaCl concentrations of 6.5–11% and sucrose concentrations of 30–50%. Ohye *et al.* (1967) reported that for four strains of *C. botulinum* type E, maximal growth rates occur at 0.995 a_w. The lowest a_w at which growth occurs is 0.965 (mixed salts solute),

TABLE 8.7.

Minimal a_w Permitting Growth of *Clostridium botulinum* Types A, B, and E at pH 7.0 and 5.5[a]

	Minimal a_w permitting growth			
	pH 7.0		pH 5.5	
Culture	Glycerol	NaCl	Glycerol	NaCl
C. botulinum A	0.93	0.96	0.95	0.97
C. botulinum B	0.93	0.96	0.95	0.97
C. botulinum E	0.95	0.98	—	0.99

[a] From Baird-Parker and Freame (1967).

and lag times are extended by as much as two- or threefold at 0.980 a_w. Scott (1955) speculated that the minimal a_w for growth of *C. botulinum* types A and B is slightly below 0.95a_w, based on studies that he conducted using canned hams. These speculations were later supported by the work of Ohye and Christian (1967), who found, using laboratory media, minimal a_w for growth levels of 0.95 for type A and 0.94 for type B (NaCl equivalents are 8.0 and 9.4%, respectively).

One of the most complete studies on the effect of a_w on *C. botulinum* growth is that of Baird-Parker and Freame (1967), who found minimal a_w levels for growth of Type A, B, and E organisms to be higher in NaCl than in glycerol (Table 8.7). In all cases, the Type E strain was much more sensitive to reduced moisture than either the Type A or B strains, and the minimal a_w levels permitting growth were influenced by pH and temperature. The observation of growth at lower a_w levels in media with added glycerol is analogous to the results reported by Kim (1965) with *C. perfringens*. The factors responsible for this "glycerol effect" are as yet unknown. Perhaps the explanation may lie in the fact that glycerol penetrates the cell rapidly, where it serves as a "compatible solute" (Brown and Simpson, 1972; Corry, 1974) and thus provides an environment that is relatively noninhibitory to the cells' metabolic processes.

The use of various factors in combination with a_w to limit the growth of *C. botulinum* is of obvious practical importance. The greatest emphasis has been placed on the interaction of nitrite and a_w, although pH and temperature have also been studied in various combinations. In studies of this type, an increase in minimal a_w levels for growth is usually observed as pH, temperatures, and/or nitrite concentrations become limiting. The data of Roberts and Ingram (1973) are excellent examples of such combination studies. Figures 8.7 and 8.8 present their findings on the interaction of $NaNO_2$, NaCl, and pH on the growth of Type B (Fig. 8.9) and Type E (Fig. 8.10) *C. botulinum* in a trypticase–peptone medium.

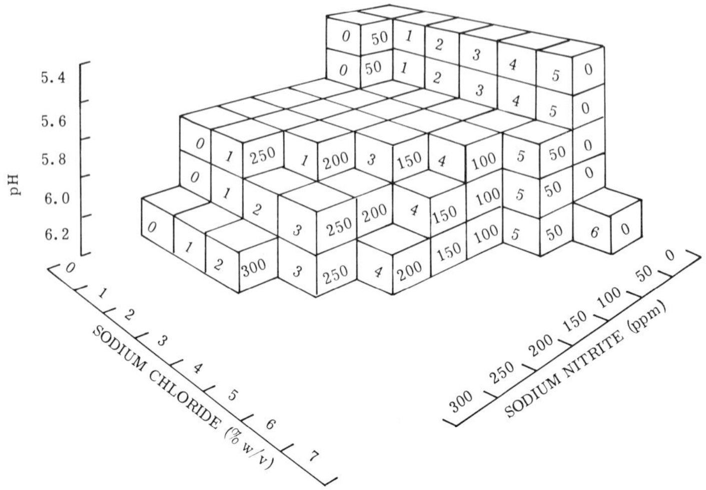

Fig. 8.9. Effects of pH, sodium chloride, and sodium nitrite on growth of *Clostridium botulinum* type B at 35°C (Roberts and Ingram, 1973).

These data indicate the greater resistance of the Type B strain to the combined effect of those factors as compared to Type E and apparent interactions, especially between NaCl level and nitrite. Yesair and Cameron (1942) observed a similar interaction between NaCl and nitrite in their studies on the germination of *C. botulinum* 62-A spores. In systems without nitrite, 5% NaCl was required to

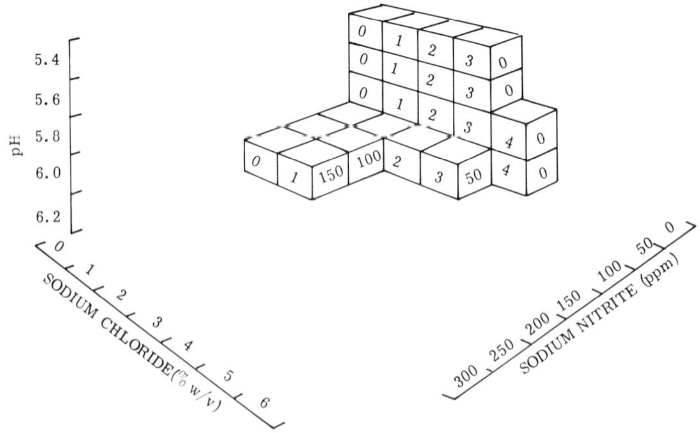

Fig. 8.10. Effects of pH, sodium chloride, and sodium nitrite on growth of *Clostridium botulinum* type E at 35°C (Roberts and Ingram, 1973).

inhibit germination, while in the presence of 0.188% $NaNO_2$, 3.5% NaCl was sufficient to produce a similar effect.

It is not yet clear whether the combined effect of various chemicals on the growth of *C. botulinum* is merely an additive inhibition of two independently inhibitory chemicals or is the result of a synergistic or interactive phenomenon. In any event, such combinations may come under even closer scrutiny in the future as health officials throughout the world search for ways to reduce the nitrite levels in foods.

Spore Germination and Sporulation

Many of the data on growth of *C. botulinum* at limited a_w levels were obtained using spore inocula and thus also characterize spore germination and outgrowth. In general, glycerol appears to support germination and outgrowth at lower a_w levels than salts, and type E strains appear to be more sensitive to limited moisture conditions than type A or B strains. Four type E strains (Schmidt and Segner, 1964) grew from spore inocula within 5–7 days in the absence of NaCl in the medium; however, outgrowth required 17–23 days in media containing 4% NaCl. These studies were performed at 10°C. At 7.8°C, 35–38 days were required for outgrowth. In another study with type E strains, Segner *et al.* (1966) found that 4.5% NaCl was required to obtain complete inhibition.

The initiation of spore germination of types A, B, and E. *C. botulinum* strains was investigated by Baird-Parker and Freame (1967), who related the commencement of growth to water activity. All strains grew from spore inocula at lower a_w levels when glycerol was the solute than when NaCl was used. These workers utilized phase-contrast microscopy to determine spore germination and observed that spores of all three types initiated germination at a_w levels down to at least 0.93 in NaCl-adjusted media. Further outgrowth did not occur within 6 weeks below 0.96 a_w in the case of types A and B, and below 0.97 a_w with type E. It was concluded that NaCl affects the sequence of events between initiation of germination and outgrowth. Glycerol, on the other hand, appears to exert its effect at levels that allow outgrowth but not vegetative cell proliferation. In addition to the type of solute used and the type of organism involved, a number of other factors influence the extent to which a_w affects spore germination and outgrowth. Type of medium, size of inoculum, temperature, pH, and the methods used to evaluate the extent of germination and outgrowth are all important in this regard.

Toxin Formation

Some early workers (Dozier, 1924, Tanner and Evans, 1933) reported toxin formation at NaCl concentrations that prevented the growth of vegetative cells of *C.*

botulinum. Whether or not growth did occur in these studies is open to some question, as these authors relied on visible turbidity as the indicator of cell multiplication. There is, thus, the possibility that a small but visually unobservable amount of growth could have produced amounts of botulinum toxin sufficient to kill the guinea pigs used as test animals. Until more complete data are available, reports indicating toxin formation in the absence of growth must be viewed with some suspicion. More recent work (Ohye and Christian, 1967) has indicated that where growth occurs, toxin formation is also demonstrable, and conversely, in the absence of growth, that cultures of *C. botulinum* are nontoxic. Minimal a_w levels supporting growth and toxin formation by types A, B, and E were 0.95, 0.94, and 0.97, respectively, which correspond to NaCl concentrations of 8.0, 9.4, and 5.1%.

In contrast, Riemann (1967) found that only 1.5–2.0% NaCl is required to inhibit toxin formation by type E strains in brain–heart infusion broth, although NaCl levels in the 4.5–5.5% range prevented growth. Working with cod homogenates, Boyd and Southcott (1971) compared three Type E strains for their ability to produce toxin, and obtained the results shown in Table 8.8. These data indicate that while the Minnesota strain appears to form toxin at slightly higher NaCl concentrations than either the Tennessee or Saratoga strains, all three variants responded in a generally similar fashion.

TABLE 8.8.

Effect of NaCl on Toxin Production in Cod Homogenates by Three Type E *Clostridium botulinum* Strains (4 Days Incubation at 30°C)[a]

Strain	NaCl (%)	Toxin, MLD[b]/gm
Minnesota	0	800
	1.73	115
	2.50	10
	2.96	10
	3.40	< 10
	3.76	0
Saratoga	0	25
	2.74	4(6 days incub.)
	3.84	0
Tennessee	0	800
	2.74	25(6 days incub.)
	3.84	0

[a] From the data of Boyd and Southcott (1971).
[b] Minimal lethal dose (mice).

From a practical standpoint, it is important to know if organoleptically acceptable levels of NaCl will prevent toxin formation and growth of *C. botulinum* in foods. Food containing a "safe" NaCl level (in the 9–11% range) would not be acceptable; thus, additional preservative factors must be present to give the same degree of safety at lower NaCl (higher a_w) levels. As noted earlier, such factors as pH, temperature, or nitrite can be used in this way. Some foods in which *C. botulinum* has grown would be expected to have objectionable odors and flavors that would render them unacceptable; however, in an interesting paper, Greenberg *et al.* (1959) showed that increasing the NaCl content of inoculated, canned meat did not suppress toxin formation until a concentration of 8.95% NaCl was reached, while abnormal odors were present only at levels up to 7.09% NaCl. Thus, the presence of 7–9% NaCl could be more hazardous than either higher or lower NaCl levels because, within this range, toxin could be synthesized without the accompanying unacceptable organoleptic changes.

Heat Resistance

The heat resistance of *C. botulinum* at various a_w levels has been studied by several workers. Weiss (1921) found that spores of this organism had increased heat resistance in canned fruit juices as the content of sucrose syrups was increased. Murrell and Scott (1966) studied the heat resistance of spores of several bacterial species, including *C. botulinum* B and E, under closely controlled a_w conditions. The greatest resistance occurred at a_w levels in the 0.2–0.4 range. Type E spores appeared to be less heat resistant than type B spores. These and other studies are discussed in the review by Troller (1973b).

The heat stability of isolated botulinal toxins as a function of a_w has not been investigated. Generally, these toxins are considered to be heat labile. Boiling for 1 minute or heating at 75°–80°C for 5–10 minutes is adequate for toxin destruction; however, heating in the presence of other protein-containing materials may stabilize the toxin to some degree. Until evidence to the contrary appears, it would be prudent to assume that reduced a_w levels protect botulinal toxins.

Much additional work is required to define many of the factors that interact with a_w reductions to prevent growth and toxin formation by *C. botulinum*. The public health importance of this organism in foods certainly requires further investigation of its toxin-producing capabilities, especially in limited moisture environments.

VIBRIO PARAHAEMOLYTICUS

Vibrio parahaemolyticus is the only known food-borne pathogen that requires NaCl for its growth. This organism inhabits the coastal waters of many countries, where it infects seafood and seafood products. Confirmed outbreaks of *V. parahaemolyticus* food poisoning have been reported in India, Taiwan, Thailand, the Philippines, Australia, and especially in Japan, where this disease causes about 70% of all cases of food-borne illness. Although the organism has been isolated in the coastal waters of the United States and implicated in several outbreaks, confirmation of the etiology of these outbreaks has been lacking. The principal reasons for the apparent localization of outbreaks to southeastern Asiatic countries are probably: (1) the extended periods for which the coastal waters maintain temperatures conducive to growth of the organism; (2) the preference of the people of these countries for dishes containing uncooked seafood; and (3) in some cases, the primitive means of sewage disposal that allow untreated wastes to enter littoral waters.

The symptoms of this disease are usually abdominal pain, nausea, and diarrhea occurring within 6–24 hours following the ingestion of contaminated food. Although deaths have been reported in Japan, it is generally believed that the mortality rate of this disease is quite low. The type of illness produced by *V. parahaemolyticus* has been described as infectious, i.e., viable organisms must be ingested for its characteristic symptoms to be produced. However, the hemolytic activity (principally to human erythrocytes—the so-called Kanagawa phenomenon) of this organism has been related to pathogenicity. The pathogenic role of the heat-stable hemolysin is supported by human feeding studies in which hemolytic strains produced disease symptoms, whereas nonhemolytic strains usually did not. Paradoxically, hemolytic strains are seldom found in seafood or in the natural marine environment, so that some process of selection or alteration may occur in the alimentary tract that converts nonhemolytic to hemolytic populations. The existence and nature of this "transformation" are as yet unknown, as is the role of the enteropathogenic substance isolated from hemolytic vibrios. Whether food-borne disease caused by *V. parahaemolyticus* is truly an infection or a toxicosis caused by the Kanagawa hemolysin or a related compound continues to be debated (Twedt and Brown, 1973).

It is only recently that the water activity requirements of *V. parahaemolyticus* have been defined. Beuchat (1974) found that the organism grows most rapidly when a laboratory medium is supplemented with 2.9% NaCl ($a_w = 0.992$), and that higher or lower a_w levels produce slower growth. The incorporation of solutes other than NaCl into tryptic soy broth to attain an a_w level of 0.992 resulted in extended lag times and generally slower rates of growth. The minimal a_w levels for four strains of *V. parahaemolyticus* with various solutes were found to be: glycerol, 0.937; KCl, 0.945; NaCl, 0.948; sucrose, 0.957; glucose, 0.983; and propylene glycol, 0.986. These results are reminiscent of those of Marshall and Scott (1958) with another vibrio. In this case, *Vibrio metchnikovi* grew very poorly in basal media of a_w above 0.998 and optimally at 0.995–0.990 a_w. At lower a_w levels, sugars were very much more inhibitory to growth than were

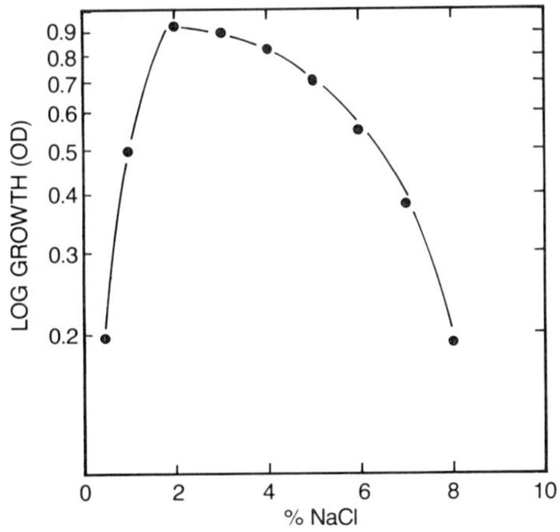

Fig. 8.11. Response of *Vibrio parahaemolyticus* growth to various concentrations of NaCl (Sakazaki, 1969).

salts. It appears that vibrios, like some halophilic bacteria, are exacting in their ionic requirements, as well as showing a preference for reduced water activities.

A report by Rödel *et al.* (1973) noted that the optimal and maximal NaCl concentrations for growth of *V. parahaemolyticus* are 3% (a_w = 0.983) and 11.0% NaCl (a_w = 0.933), respectively. These results agree with the data of Sakazaki (1969) given in Fig. 8.11, which show a sharp decline in growth at NaCl concentrations of less than 2%, with a slower decline in growth rates also appearing between 3 and 8%. A property used in differentiating this species from other vibrios is growth in the presence of 8% but not 10% NaCl.

Although glycerol in low concentrations was found by Beuchat (1974) to be the least inhibitory to growth of *V. parahaemolyticus* of a number of solutes tested, Chun *et al.* (1972) found that a buffered salt solution containing 30% glycerol (which has been recommended for transporting enteric pathogens) is rapidly lethal to this organism. Similar intolerance to high glycerol concentrations has been noted for a number of other enteric pathogens. The ability of *V. parahaemolyticus* to survive heating with NaCl at various concentrations was tested by Covert and Woodburn (1972). The concentrations tested (3–12%) all increased survival of the organism when heated in trypticase–soy broth at 48°C. Survival during low temperature storage of the cells (−5° and −18°C) was also enhanced by added NaCl. Similar studies of both high and low temperature survival in a fish homogenate produced essentially parallel results, with the homogenate itself providing some additional protection over the range of NaCl concentrations used. A similar effect was noted by Beuchat and Worthington (1976), who found that D_{47° values are increased as the NaCl content of the heating menstruum was increased. Maximal resistance occurred in a phosphate buffer containing 7.5% NaCl, the highest NaCl concentration tested.

Considerable speculation continues to surround the importance of *V. parahaemolyticus* as a food-borne disease organism in foods other than seafoods. Thus far, there have been no such reported and documented outbreaks; however, at least one study (Nelson and Potter, 1976) showed that growth of this organism does occur in certain proteinaceous, laboratory-prepared foods. If the minimal NaCl level required for growth is indeed 3%, it would seem that growth in other than marine-related environments would be unlikely. On the other hand, the above report indicates that growth occurs at NaCl concentrations as low as 0.3% (0.0513 *M*) and that K^+ or Li^+ can be substituted for Na^+. These authors speculate that this organism can utilize cations in an additive fashion to fulfill their ionic requirements. If further experimentation should confirm these speculations, it would seem possible that nonmarine foods may well serve as a vehicle for this organism. Perhaps the lack of confirmatory, epidemiological evidence in this regard simply reflects the fact that the organisms have not been looked for in foods other than seafoods.

Obviously, additional work is needed to establish the ionic environment required for the growth and survival of *V. parahaemolyticus*. The implications of such findings undoubtedly will reveal the hazard that this organism poses in all foodstuffs.

BACILLUS CEREUS

Food-borne illnesses caused by *B. cereus* seem to be characterized by two different patterns of symptomatology. The first is associated with relatively mild symptoms involving intestinal discomfort such as abdominal cramps and diarrhea. The onset of these symptoms is normally 6–16 hours after ingestion of the contaminated food. This type of disease has been regarded as the "classic" type of *B. cereus* food poisoning and has resulted in 10 or 12 reported outbreaks in the United States between 1968 and 1974. In Hungary, 8.2% of all outbreaks of food-borne diseases are caused by *B. cereus*, with meat or meat-containing dishes being most frequently implicated. Other European countries, such as the Netherlands, Sweden, and Norway, also have reported a number of cases of this disease, with a wide variety of foods involved.

Recently, a more severe form of the disease has been encountered in the United Kingdom often related to the consumption of cooked rice. In these cases, the onset of symptoms, usually characterized by vomiting and nausea, occurs within 1–6 hours after consumption of the implicated food. Details of this disease can be found in the reports of Goepfert *et al.* (1972), Gilbert *et al.* (1974), Hobbs (1974), and Taylor and Gilbert (1975).

The reason for this diverse symptomatology is not apparent at present; however, it is suspected that different toxins are responsible for the two forms of the disease. Research on the serotyping of flagellar antigens (Taylor and Gilbert, 1975) appears to be making some progress in identifying the emetic strains, although many are untypable. Serological studies also have revealed that the organisms responsible for the emetic and the milder, diarrheic forms are distinguishable serologically.

Usually, *B. cereus* food poisoning occurs when a food is heated insufficiently to kill *B. cereus* spores or to inactivate preformed enterotoxin. Subsequent holding of the food at temperatures favorable to spore germination and outgrowth may complete the requirements for rapid proliferation of the organism and enterotoxin production.

Despite the fact that NaCl is a common ingredient in the differential media used to characterize *B. cereus,* relatively little has been published of its water requirements. The principal reference on the effects of a_w on germination and growth of *B. cereus* is the paper of Jakobsen *et al.* (1972). These authors, as well as Scott (1957) and Raevuori and Genigeorgis (1975), agree that the minimal a_w for growth in NaCl-containing media is 0.95. This minimum may be somewhat lower in the presence of other solutes, however. Minimal values for spore germination and vegetative cell division were determined by these authors at various a_w levels obtained with a variety of solutes. Nearly all spores were found to germinate at 0.99–0.98 a_w in the presence of NaCl and KCl, but decreasing a_w reduced germination rates with total inhibition occurring below 0.95 a_w. Inhibition of vegetative cell growth (Table 8.9) also occurred, depending to some extent on the nature of the solute used. The solute that allowed growth and germination at the lowest a_w levels was glycerol, with dimethylsulfoxide (DMSO) and erythritol somewhat more inhibitory. Obviously, the intrinsic properties of the solutes influenced the minimal a_w levels for germination and spore outgrowth. These authors speculated that the greater ability of *B. cereus* to grow and germinate in solutions of erythritol, DMSO, and glycerol might be related to the capacity of these compounds to form hydrogen bonds and thus replace intracellular water.

The above report indicating that an a_w level of 0.95 (8% NaCl) prevents germination of *B. cereus* spores is supported by the data of Gould (1964), who found that slightly higher NaCl concentrations (10–15%, a_w = 0.94–0.91 estimated) are required to prevent germination of spores obtained from a number of species of the genus *Bacillus*. Generally, however, outgrowth of spores is more sensitive to increasing NaCl concentrations than germination and is inhibited in the 6–7% NaCl range.

Beyond the observation that cured, brined, or syruped foods have not been implicated in outbreaks of *B. cereus* food poisoning, there is little information on the growth of this organism in foods. Its water relations are similar to those of the salmonellae and, clearly, it will not be a hazard in foods of relatively low a_w

TABLE 8.9.

Minimal a_w Levels for Germination and Outgrowth of *B. cereus* Spores[a]

	a_w Levels with observed germination during 14 days incubation		Minimal a_w permitting cell division during 14 days incubation
Solute	> 90%	> 10%	
NaCl	0.98	0.95	0.95
KCl	0.98	0.95	0.95
Fructose	0.98	0.95	0.97
Glucose	0.97	0.94	0.95
Sorbitol	0.97	0.94	0.94
Erythritol	0.94	< 0.90	0.94
Glycerol	0.92	< 0.85	0.93
DMSO	0.94	< 0.90	0.94

[a] Data from Jakobsen *et al.* (1972).

(≤ 0.90), as are the staphylococci. *B. cereus* sporulates freely in foods and causes food poisoning most often in foods that have been heated. The data of Davies and Wilkinson (1973) suggest that a heat shock is important in stimulating germination of this species. In practice, this will both initiate growth in a food and reduce the number of competing bacteria.

PARASITES

Caused by the nematode, *Trichinella spiralis,* trichinosis is one of the most common food-related parasitic infections. Insufficiently cooked pork is the major source of this disease in man, although the consumption of meat from game animals, especially bear, also may lead to infection.

Usually, thorough cooking, that is, with all parts of the meat reaching a temperature of at least 58°C, is sufficient to inactivate the larvae encapsulated within the muscle tissue. Storage of meat at subfreezing temperatures for at least 20 to 30 days, depending on the storage temperature, also will kill trichinae.

The use of uncooked and potentially contaminated pork in cured sausages has resulted in the establishment of regulatory controls in many countries with respect to the time and temperature of smoking and the concentration of NaCl in the meat. For example, in West Germany, dry imported sausage need not be inspected for trichinae only if this product contains at least 4% NaCl and possesses a water content of < 25% (Lötsch and Rödel, 1974). The intent of such regulations is to assure that all trichinal cysts are killed during the curing process.

The relationship between the viability of encysted *Trichina* and the salt content of curing brines and, ultimately, the meat, is poorly defined. Crouse and Kemp (1969) showed that viable trichinae persisted in hams and pork shoulders during curing (brine containing approximately 6% NaCl), salt equalization, smoking, and throughout the initial two weeks of aging at 75°F. Following three weeks of aging, however, the meat tissues of both products were found to be free of parasites. Zimmerman (1971) reported that temperature rather than salt concentration was the primary critical factor in killing encysted trichinae in hams; however, the importance of salt in the devitalization of these parasites also was stressed. This toxicity also was demonstrated by Ransom *et al.* (1920) and by Allen and Goldberg (1962). Whether or not the effect is solute specific, that is, specific only to NaCl, or is a result of a_w limitation has not been demonstrated.

Lötszch and Rödel (1974) have reported that for sausages at an a_w level of 0.90, one should not expect to find trichinae in dry sausage; however, these a_w limits are based on the addition of NaCl to sausages. The effect of other humectants on trichinal survival in hams and similar products has not been reported. Similarly, the interaction of a_w effects with the components of smoke, including various aldehydes, ketones, cresols, and phenolic substances, has not been investigated.

The effect of reduced a_w levels on the agents of food-borne parasitic infections other than those caused by *Trichinella spiralis* has not been reported.

These data on the influence of water activity on the growth, survival, and where applicable, toxin formation of food-borne pathogens, by no means cover all of the microorganisms that have been described as causing food illnesses. For example, little or no data exist on the effect of a_w on the survival of viruses found in foods, although moisture conditions must be important. Other "occasional" food-borne pathogens, such as *Pseudomonas* and *Escherichia coli,* will be dealt with elsewhere in this volume. *Shigella* species are more commonly thought to be water-borne agents of disease, and thus the moisture requirements of this genus are not considered here. It is probable, however, that the water requirements of shigellae would be similar to those of salmonellae.

REFERENCES

Allen, R. W., and Goldberg, A. (1962). The effect of various salt concentrations on encysted *Trichinella spiralis* larvae. *Am. J. Vet. Res.* **23,** 580–586.

Ayerst, G. (1969). The effects of moisture and temperature on growth and spore germination in some fungi. *J. Stored Prod. Res.* **5,** 127–141.

Bacon, C. W., Sweeney, V. G., Robbins, J. D., and Burdick, D. (1973). Production of penicillic acid and ochratoxin A on poultry feed by *Aspergillus ochraceus:* Temperature and moisture requirements. *Appl. Microbiol.* **26,** 155–160.

Baer, E. F., Franklin, M. K., and Gilden, M. M. (1966). Efficiency of several selective media for isolating coagulase-positive staphylococci from food products. *J. Assoc. Off. Anal. Chem.* **49,** 267–269.

Baird-Parker, A. C. (1962). An improved diagnostic and selective medium for isolating coagulase-positive staphylococci. *J. Appl. Bacteriol.* **25,** 12–19.

Baird-Parker, A. C. (1971). Factors affecting production of bacterial food poisoning toxins. *J. Appl. Bacteriol.* **34,** 181–197.

Baird-Parker, A. C., and Davenport, E. (1965). The effect of recovery medium on the isolation of *Staphylococcus aureus* after heat treatment and after the storage of frozen or dried cells. *J. Appl. Bacteriol.* **28,** 390–402.

Baird-Parker, A. C., and Freame, B. (1967). Combined effect of water activity, pH and temperature on the growth of *Clostridium botulinum* from spore and vegetative cell inocula. *J. Appl. Bacteriol.* **30,** 420–429.

Baird-Parker, A. C., Boothroyd, M., and Jones, E. (1970). The effect of water activity on the heat resistance of heat sensitive and heat resistant strains of salmonellae. *J. Appl. Bacteriol.* **33,** 515–522.

Bean, P. G., and Roberts, T. A. (1975). Effect of sodium chloride and sodium nitrite on the heat resistance of *Staphylococcus aureus* NCTC 10652 in buffer and meat macerate. *J. Food Technol.* **10,** 327–332.

Beuchat, L. R. (1974). Combined effects of water activity, solute, and temperature on the growth of *Vibrio parahaemolyticus. Appl. Microbiol.* **27,** 1075–1080.

Beuchat, L. R., and Worthington, R. E. (1976). Relationships between heat resistance and phospholipid fatty acid composition of *Vibrio parahaemolyticus. Appl. Environ. Microbiol.* **31,** 389–394.

Blanche Koelensmid, W. A. A., and van Rhee, R. (1964). Salmonella in meat products. *Ann. Inst. Pasteur Lille* **15,** 85–97.

Boyd, J. W., and Southcott, B. A. (1971). Effects of sodium chloride on outgrowth and toxin production of *Clostridium botulinum* Type E in cod homogenates. *J. Fish. Res. Board Can.* **28,** 1071–1075.

Boylan, S. L., Acott, K. A., and Labuza, T. P. (1976). *Staphylococcus aureus* challenge study in an intermediate moisture food. *J. Food Sci.* **41,** 918–921.

Brown, A. D., and Simpson, J. R. (1972). Water relations of sugar-tolerant yeasts: The role of intracellular polyols. *J. Gen. Microbiol.* **72,** 589–591.

Calhoun, C. L., and Frazier, W. C. (1966). Effects of available water on thermal resistance of three nonsporeforming species of bacteria. *Appl. Microbiol.* **14,** 416–420.

Chapman, G. H. (1945). The significance of sodium chloride in studies of staphylococci *J. Bacteriol.* **50,** 201–203.

Chapman, G. H. (1948). An improved medium for the isolation and testing of food-poisoning staphylococci. *Food Res.* **13,** 100–105.

Chordash, R. A., and Potter, N. N. (1972). Effects of dehydration through the intermediate moisture range on water activity, microbial growth, and texture of selected foods. *J. Milk Food Technol* **35,** 395–398.

Christian, J. H. B. (1955). The influence of nutrition on the water relations of *Salmonella oranienburg* at 30°C. *Aust. J. Biol. Sci.* **8,** 490–497.

Christian, J. H. B., and Hall, J. M. (1972). Water relations of *Salmonella oranienburg:* Accumulation of potassium and amino acids during respiration. *J. Gen. Microbiol.* **70,** 497–506.

Christian, J. H. B., and Scott, W. J. (1953). Water relations of salmonellae at 30°C. *Aust. J. Biol. Sci.* **6,** 565–573.

Christian, J. H. B., and Stewart, B. J. (1973). Survival of *Staphylococcus aureus* and *Salmonella newport* in dried foods, as influenced by water activity and oxygen. *Microbiol. Saf. Food, Proc. Int. Symp. Food Microbiol. 8th, 1972* pp. 107–119.

Christian, J. H. B., and Waltho, J. A. (1966). Water relations of *Salmonella oranienburg:* Stimulation of respiration by amino acids. *J. Gen. Microbiol.* **43,** 345–355.

Chun, D., Seol, S. Y., Tak, R., and Park, C. K. (1972). Inhibitory effect of glycerin on *Vibrio parahaemolyticus* and *Salmonella*. *Appl. Microbiol.* **24,** 675–678.

Clayson, D. H. F., and Blood, R. M. (1957). Food perishability: The determination of the vulnerability of food surfaces to bacterial infection. *J. Sci. Food Agric.* **8,** 404–414.

Collins-Thompson, D. L., Hurst, A., and Aris, B. (1974). Comparison of selective media for the enumeration of sublethally heated food-poisoning strains of *Staphylococcus aureus*. *Can. J. Microbiol.* **20,** 1072–1075.

Communicable Disease Center. (1974). Annual Report. U.S. Dept. of Health, Education, and Welfare, Public Health Serv., Atlanta, Georgia.

Corry, J. E. L. (1974). The effect of sugars and polyols on the heat resistance of salmonellae. *J. Appl. Bacteriol.* **37,** 31–43.

Corry, J. E. L. (1976). Sugar and polyol permeability of *Salmonella* and osmophilic yeast cell membranes measured by turbidimetry, and its relation to heat resistance. *J. Appl. Bacteriol.* **40,** 277–284.

Covert, D., and Woodburn, M. (1972). Relationships of temperature and sodium chloride concentration to the survival of *Vibrio parahaemolyticus* in broth and fish homogenate. *Appl. Microbiol.* **23,** 321–325.

Crouse, J. D., and Kemp, J. D. (1969). Salt and aging time effects on the viability of *Trichinella spiralis* in heavy dry-cured hams and shoulders. *J. Food Sci.* **34,** 530–521.

Czop, J. K., and Bergdoll, M. S. (1974). Staphylococcal enterotoxin synthesis during the exponential, transitional and stationary growth phases. *Infect. Immun.* **9,** 229–235.

Dack, G. M., Cary, W. E., Woolpert, O., and Wiggers, H. (1930). An outbreak of food poisoning proved to be due to yellow hemolytic staphylococci. *J. Prev. Med.* **4,** 167–175.

Davies, F. L., and Wilkinson, G. (1973). *Bacillus cereus* in milk and dairy products. *Microbiol. Saf. Food, Proc. Int. Symp. Food Microbiol., 8th, 1972* pp. 57–67.

Diener, U. L., and Davis, N. D. (1967). Limiting temperature and relative humidity for growth and production of aflatoxin and free fatty acids by *Aspergillus flavus* in sterile peanuts. *J. Am. Oil Chem. Soc.* **44,** 259–263.

Diener, U. L., and Davis, N. D. (1970). Limiting temperature and relative humidity for aflatoxin production by *Aspergillus flavus* in stored peanuts. *J. Am. Oil Chem. Soc.* **47,** 347–351.

Doyle, M. P., and Marth, E. H. (1975). Thermal inactivation of conidia from *Aspergillus flavus* and *Aspergillus parasiticus*. II. Effects of pH and buffers, glucose, sucrose, and sodium chloride. *J. Milk Food Technol.* **38,** 750–758.

Dozier, C. C. (1924). Inhibitive influences of sugars and salts on viability, growth, and toxin production of *Bacillus botulinum*. *J. Infect. Dis.* **35,** 134–155.

Foster, E. M., Goepfert, J. M., and Deibel, R. H. (1970). Detecting presence of *Salmonella:* Rapid methods. *Manuf. Confect.* **50,** 57–60.

Garibaldi, J. A. (1968). Acetic acid as a means of lowering the heat resistance of *Salmonella* in yolk products. *Food Technol.* **22,** 1031–1033.

Genigeorgis, C., Riemann, H., and Sadler, W. W. (1969). Production of enterotoxin B in cured meats. *J. Food Sci.* **34,** 62–68.

Genigeorgis, C., Foda, M. S., Mantis, A., and Sadler, W. W. (1971). Effect of sodium chloride and pH on enterotoxin C production. *Appl. Microbiol.* **21,** 862–866.

Gibson, B. (1973). The effect of high sugar concentrations on the heat resistance of vegetative microorganisms. *J. Appl. Bacteriol.* **36,** 365–376.

Gilbert, R. A., Stringer, M. F., and Peace, T. C. (1974). The survival and growth of *Bacillus cereus* in boiled and fried rice in relation to outbreaks of food poisoning. *J. Hyg.* **73,** 433–444.

Goepfert, J. M., Iskander, I. K., and Amundson, C. H. (1970). Relation of the heat resistance of salmonellae to the water activity of the environment. *Appl. Microbiol.* **19,** 429–433.

Goepfert, J. M., Spira, W. N., and Kim, H. U. (1972). *Bacillus cereus:* Food poisoning organism. A review. *J. Milk Food Technol.* **35,** 213–227.

Gough, B. J., and Alford, J. A. (1965). Effect of curing agents on the growth and survival of food-poisoning strains of *Clostridium perfringens*. *J. Food Sci.* **30,** 1025–1028.

Gould, G. W. (1964). Effect of food preservatives on the growth of bacteria from spores. *Proc. Int. Symp. Food Microbiol., 4th, 1963* pp. 17–24.

Greenberg, R. A., Silliker, J. H., and Fatta, L. D. (1959). The influence of sodium chloride on toxin production and organoleptic breakdown in perishable cured meat inoculated with *Clostridium botulinum*. *Food Technol.* **13,** 509–511.

Halls, N. A., and Ayres, J. C. (1973). Potential production of sterigmatocystin on country-cured ham. *Appl. Microbiol.* **26,** 636–637.

Harwig, J., and Chen. Y. K. (1974). Some conditions favoring production of ochratoxin A and citrinin by *Penicillium viridicatum* in wheat and barley. *Can. J. Plant Sci.* **54,** 17–22.

Hauschild, A. H. W. (1971). *Clostridium perfringens* enterotoxin. *J. Milk Food Technol.* **34,** 596–599.

Heidelbauer, R. J., and Middaugh, P. R. (1973). Inhibition of staphylococcal enterotoxin production in convenience foods. *J. Food Sci.* **38,** 885–888.

Hill, J. H., and White, E. C. (1929). The use of sodium chloride in culture media for the separation of certain gram-positive cocci from gram-negative bacilli. *J. Bacteriol.* **1,** 47.

Hobbs, B. (1974). *Clostridium welchii* and *Bacillus cereus* infection and intoxication. *Postgrad. Med. J.* **50,** 597–602.

Hojvat, S. A., and Jackson, H. (1969). Effects of sodium chloride and temperature on the growth and production of enterotoxin B by *Staphylococcus aureus*. *Can. Inst. Food Technol. J.* **2,** 56–59.

Hsieh, F., Acott, K., Elizondo, H., and Labuza, T. P. (1975). The effect of water activity on the heat resistance of vegetative cells in the intermediate moisture range. *Lebensm.-Wiss. u. Technol.* **8,** 78–81.

Incze, K., and Frank, H. K. (1976). Is there a danger of mycotoxins in Hungarian sausage? Effect of substrate a_w value and temperature on toxin production in mixed cultures. *Fleischwirtschaft* **56,** 219–225.

Jakobsen, M., Filtenberg, O. L., and Bramsnaes, F. (1972). Germination and outgrowth of the bacterial spore in the presence of different solutes. *Lebensm.-Wiss. u. Technol.* **5,** 159–162.

Jarvis, B. (1971). Factors affecting the production of mycotoxins. *J. Appl. Bacteriol.* **34,** 199–213.

Kadan, R. S., Martin, W. H., and Mickelsen, R. (1963). Effects of ingredients used in condensed and frozen dairy products on thermal resistance of potentially pathogenic staphylococci. *Appl. Microbiol.* **11,** 45–49.

Kang, C. K., Woodburn, M., Pagenkopf, A., and Cheney, R. (1969). Growth, sporulation, and germination of *Clostridium perfringens* in media of controlled water activity. *Appl. Microbiol.* **18,** 798–805.

Kelly, F. C., and Dack, G. M. (1936). Experimental food poisoning. A study of a food poisoning staphylococcus and the production of an enterotoxic substance in bread and meat. *Am. J. Public Health* **26,** 1077–1082.

Kim, C. H. (1965). Substrate factors for growth and sporulation of *Clostridium perfringens* in selected foods and in simple systems. Ph.D. Thesis, Purdue University, Lafayette, Indiana (University Microfilms No. 65-8624).

King, W. L., and Hurst, A. (1963). A note on the survival of some bacteria in different diluents. *J. Appl. Bacteriol.* **26,** 504–506.

Koch, F. E. (1942). Selective media for staphylococci. *Zentralbl. Bakteriol., Parasitenkd., Infektionski. Hyg., Abt. 1: Orig.* **149,** 122–124.

Lechowich, R. V., Evans, J. B., and Niven, C. F. (1956). Effect of curing ingredients and procedures on the survival and growth of staphylococci in and on cured meats. *Appl. Microbiol.* **4,** 360–363.

Liu, T. S., Snoeyenbos, G. H., and Carlson, V. L. (1969). The effect of moisture and storage temperature on a *Salmonella senftenberg* 775 W population in meat and bone meal. *Poul. Sci.* **48,** 1628–1633.

Lötsch, R., and Rödel, W. (1974). Investigations into the viability of *Trichinella spiralis* in dry sausages as a function of water activity. *Fleischwirtschaft* **54,** 1203–1208.

McLean, R. A., Lilly, H. D., and Alford, J. A. (1968). Effects of meat curing salts and temperature on production of staphylococcal enterotoxin B. *J. Bacteriol.* **95,** 1207–1211.

McDonough, F. E., and Hargrove, R. E. (1968). Heat resistance of *Salmonella* in dried milk. *J. Dairy Sci.* **51,** 1587–1591.

Marshall, B. J., and Scott, W. J. (1958). The water relations of *Vibrio metchnikovi* at 30 C. *Aust. J. Biol. Sci.* **11,** 171–176.

Marshall, B. J., Ohye, D. F., and Christian, J. H. B. (1971). Tolerance of bacteria to high concentrations of NaCl and glycerol in the growth medium. *Appl. Microbiol.* **21,** 363–364.

Mead, G. C. (1969). Combined effect of salt concentration and redox potential of the medium on the initiation of vegetative growth by *Clostridium welchii*. *J. Appl. Bacteriol.* **32,** 468–475.

Mossel, D. A. A., and Koopman, M. J. (1965). Losses in viable cells of salmonellae upon inoculation into dry animal feeds of various types. *Poult. Sci.* **44,** 890–893.

Murrell, W. G., and Scott, W. J. (1966). The heat resistance of bacterial spores at various water activities. *J. Gen. Microbiol.* **43,** 411–425.

Nelson, K. J., and Potter, N. N. (1976). Growth of *Vibrio parahemolyticus* at low salt levels and in nonmarine foods. *J. Food Sci.* **41,** 1413–1417.

Niskanen, A., and Nurmi, E. (1976). Effect of starter culture on staphylococcal enterotoxin and thermonuclease production in dry sausage. *Appl. Environ. Microbiol.* **31,** 11–20.

Northolt, M. D., Verhulsdonk, C. A. H., Soentoro, P. S. S., and Paulsch, W. E. (1976). Effect of water activity and temperature on aflatoxin production by *Aspergillus parasiticus*. *J. Milk Food Technol.* **39,** 170–174.

Nunheimer, T. D., and Fabian, F. W. (1940). Influence of organic acids, sugars and sodium chloride upon strains of food poisoning staphylococci. *Am. J. Public Health* **30,** 1040–1048.

Ohye, D. F., and Christian, J. H. B. (1967). Combined effects of temperature pH and water activity on growth and toxin production of *Cl. botulinum* Types A, B, and E. *Botulism 1966, Proc. Int. Symp. Food Microbiol., 5th, 1966* pp. 217–223.

Ohye, D. F., Christian, J. H. B., and Scott, W. J. (1967). Influence of temperature on the water relations of growth of *Cl. botulinum* Type E. *Botulism 1966, Proc. Int. Symp. Food Microbiol., 5th, 1966* pp. 136–143.

Orth, R. (1976). The influence of water activity on the spore germination of aflatoxin, sterigmatocystin, and patulin producing molds. *Lebensm.-Wiss. u. Technol.* **9,** 156–159.

Oxhøj, P. (1943). The influence of NaCl on the growth of certain anaerobic bacteria. *Arsskr., K. Vet.* Landsbohoejsk., Copenhagen pp. 1–45.

Raevuori, M., and Genigeorgis, C. (1975). Effect of pH and sodium chloride on growth of *Bacillus cereus* in laboratory media and certain foods. *Appl. Microbiol.* **29,** 68–73.

Ransom, B. H., Schwartz, B., and Raffensperger, H. H. (1920). Effects of post-curing processes on trichinae. U.S.D.A. Bull. 880, 37 pp.

Riemann, H. (1967). The effect of the number of spores on growth and toxin formation by *Cl. botulinum* Type E in inhibitory environments. *Botulism 1966, Proc. Int. Symp. Food Microbiol., 5th, 1966* pp. 150–157.

Riemann, H. (1968). Effect of water activity on the heat resistance of *Salmonella* in "dry" materials. *Appl. Microbiol.* **16,** 1621–1622.

Roberts, T. A., and Ingram, M. (1973). Inhibition of growth of *Cl. botulinum* at different pH values by sodium chloride and sodium nitrite. *J. Food Technol.* **8,** 467–475.

Rödel, W., Herzog, H., and Leistner, L. (1973). Wasseraktivitäts-Toleranz von lebensmit-

telhygienisch wichtigen Keimerten der Gattung Vibrio. *Fleischwirtschaft* **53,** 1301–1303.
Sakazaki, R. (1969). Halophilic vibrio infections. *In* "Foodborne Infections and Intoxications" (H. Riemann, ed.), pp. 115–129. Academic Press, New York.
Schmidt, C. F., and Segner, W. P. (1964). The bacteriology of Type E *Clostridium botulinum*. *Proc. Res. Conf. Res. Counc. Am. Meat Inst. Found., Univ. Chicago* **16,** 13–20.
Scott, W. J. (1953). Water relations of *Staphylococcus aureus* at 30 C. *Aust. J. Biol. Sci.* **6,** 549–563.
Scott, W. J. (1955). Factors in canned ham controlling *Cl. botulinum* and *Staph. aureus*. *Ann. Inst. Pasteur Lille* **7,** 68–73.
Scott, W. J. (1957). Water relations of food spoilage microorganisms. *Adv. Food Res.* **7,** 83–127.
Segalove, M., and Dack, G. M. (1951). Growth of bacteria associated with food poisoning experimentally inoculated into dehydrated meat. *Food Res.* **16,** 118–125.
Segner, W. P., Schmidt, C. F., and Boltz, J. K. (1966). Effect of sodium chloride and pH on the outgrowth of spores of Type E *Clostridium botulinum* at optimal and suboptimal temperatures. *Appl. Microbiol.* **14,** 49–54.
Shih, C. N., and Marth, E. H. (1972). Production of aflatoxin in a medium fortified with sodium chloride. *J. Dairy Sci.* **55,** 1415–1419.
Spencer, R. (1969). Food poisoning due to clostridia. The factors affecting the survival and growth of food poisoning clostridia in cured foods. *B.F.M.I.R.A. Sci. Tech. Surv.* No. 58.
Spira, W. M., and Goepfert, J. M. (1972). *Bacillus cereus*-induced fluid accumulation in rabbit ileal loops. *Appl. Microbiol.* **24,** 341–348.
Stark, R. L., and Duncan, C. L. (1972). Purification and biochemical properties of *Clostridium perfringens* Type A enterotoxin. *Infect. Immun.* **6,** 662–673.
Stone, R. V. (1935). A cultural method for classifying staphylococci as of the "food poisoning" type. *Proc. Soc. Exp. Biol. Med.* **33,** 185–187.
Stott, W. T., and Bullerman, L. B. (1975). Instability of patulin in cheddar cheese. *J. Food Sci.* **41,** 201–203.
Strong, D. H., Foster, E. M., and Duncan, C. L. (1970). Influence of water activity on the growth of *Clostridium perfringens*. *Appl. Microbiol.* **19,** 980–987.
Tanner, F. W., and Evans, F. I. (1933). Effect of meat curing solutions on anaerobic bacteria. I. Sodium chloride. *Zentralbl. Bakteriol., Parasitenkd., Infektionskr. Hyg., Abt. 2* **88,** 44–54.
Taylor, A. J., and Gilbert, R. C. (1975). *Bacillus cereus* food poisoning: A provisional serotyping scheme. *J. Med. Microbiol.* **8,** 543–550.
Troller, J. A. (1971). Effect of water activity on enterotoxin B production and growth of *Staphylococcus aureus*. *Appl. Microbiol.* **21,** 435–439.
Troller, J. A. (1972). Effect of water activity on enterotoxin A production and growth of *Staphylococcus aureus*. *Appl. Microbiol.* **24,** 440–443.
Troller, J. A. (1973a). Effect of water activity and pH on staphylococcal enterotoxin B production. *Acta Aliment. Acad. Sci. Hung.* **2,** 351–360.
Troller, J. A. (1973b). The water relations of food-borne bacterial pathogens. A review. *J. Milk Food Technol.* **36,** 276–288.
Troller, J. A. (1975). Influence of water activity on growth and enterotoxin formation by *Staphylococcus aureus* in foods. *J. Food Sci.* **40,** 802–804.
Troller, J. A., and Frazier, W. C. (1963). Repression of *Staphylococcus aureus* by food bacteria. I. Effect of environmental factors on inhibition. *Appl. Microbiol.* **11,** 11–14.
Twedt, R. M., and Brown, D. F. (1973). *Vibrio parahaemolyticus:* Infection or toxicosis? *J. Milk Food Technol.* **36,** 129–134.
Van Walbeek, W., Scott, P. M., and Thatcher, F. S. (1968). Mycotoxins from foodborne fungi. *Can. J. Microbiol.* **14,** 131–137.
Weiss, H. (1921). The thermal death point of the spores of *Bacillus botulinus* in canned foods. *J. Infect. Dis.* **29,** 362–368.

Yesair, J., and Cameron, E. J. (1942). Inhibitive effect of curing agents on anaerobic spores. *Canner* **94,** 89–92.

Young, J. R. (1967). Humidity control in the laboratory using salt solutions—a review. *J. Appl. Chem.* **17,** 241–245.

Zimmerman, W. J. (1971). Salt cure and drying-time and temperature effects on viability of *Trichinella spiralis* in dry-cured hams. *J. Food Sci.* **36,** 58–62.

9

Control of a_w and Moisture

Controlling the a_w and/or moisture content of foods is a broad subject covering diverse principles, processes, and products. The intent of this chapter is to characterize some of these processes from an "a_w point of view." Moisture removal from foods will influence a_w to a greater or lesser degree, and a discussion of how a_w can be reduced in foods is therefore considered appropriate to this volume. For a more extensive and theoretical development of this subject, the reader should consult Van Arsdale *et al.* (1973), Karel (1973), and Keey (1972). Our emphasis in this chapter will be on reviewing the various methods of moisture and a_w reduction in foods and the implications that a_w has for these techniques.

DEHYDRATION

The preservation of foods by drying is an ancient practice that allowed early man to survive in times of food scarcity and to separate himself, at least temporarily, from his food sources during periods of migration. It is a widespread art practiced in pre-Columbian, ancient Eskimo, Mediterranean, Oriental, and African cultures. With the passage of time, the sun drying techniques employed by these early cultures were often supplanted by drying in crude ovens and similar devices. As is frequently the case, more modern approaches to preserving foods, in this instance by commercial dehydration, have been developed by nations and governments eager to maintain far-flung military operations. Military needs have continued to provide much of the impetus to the development of dehydration technology, although commercial and nonmilitary governmental agencies have led the way in many of the most recent advances.

In terms of world production, dehydrated (usually spray-dried) skim milk probably surpasses in quantity all other artificially or mechanically dried processed foods. Figs, raisins, and prunes are the leading dried fruits, and potatoes and onions are probably the principal dehydrated vegetables.

Basically, drying is a process in which water is transferred from a food to a gas, or a gas mixture (usually air). The gas is then allowed to escape to the atmosphere or it may be recirculated if some portion of the water vapor that it contains has been removed. This concept has been extended by the substitution of a water-absorbing material for the gas. If sufficient time is allowed for equilibration between the solute and the food to occur, an effective reduction in a_w may be achieved.

Although simple in concept, drying is a highly complex physical phenomenon in which energy, usually in the form of heat, is supplied to the water contained in foods (Keey, 1972). This heat raises the vapor pressure of the water to such an extent that it evaporates from the food surface. Food surface water thus removed is replaced with water from within by internal moisture transfer, a term used to describe a number of processes such as diffusion, convection, capillary flow, physical deformation of the food (shrinkage), and other means. The increase in vapor pressure noted above is heat-requiring (endothermic), and so heat is lost—a phenomenon known as evaporative cooling. Thus, to maintain the temperature of the food surface sufficient to raise the water vapor pressure to a point at which water is removed, heat must be continually provided. In essence, the force driving this transfer is the vapor pressure difference between the atmosphere "receiving" the water and the evaporating surface that "contributes" the water. The relationship between vapor pressure difference and temperature is expressed in the formula [Eq. (1)] of Krischer (1939):

$$P_{sw} - P_w = \frac{P(t_a - t_w)}{2720} \qquad (1)$$

where P_{sw} is the vapor pressure of water at the wet-bulb temperature, P_w is the partial pressure of water vapor in the air, P is the barometric pressure, t_a is the air temperature, and t_w is the wet-bulb temperature.

If moisture loss from a food is plotted against time, the curve formed is biphasic (Fig. 9.1). The initial phase is characterized by a drying rate that is constant and independent of water content. The water removed during this phase behaves as if it were a continuous film. In the case of nonporous materials, this film represents superficial water, whereas in a porous solid, the water removed from the surface must be supplied from the interior of the material.

As moisture continues to be removed, the constant-rate period changes, usually rather abruptly, into the falling-rate phase in which the drying rate decreases with decrease in moisture content. This point is often described as the critical moisture content, and in solid foods (porous materials) it is reached when the moisture flow to the surface is no longer equal to the rate of evaporation. In the falling-rate phase, restrictive forces are present in addition to the internal

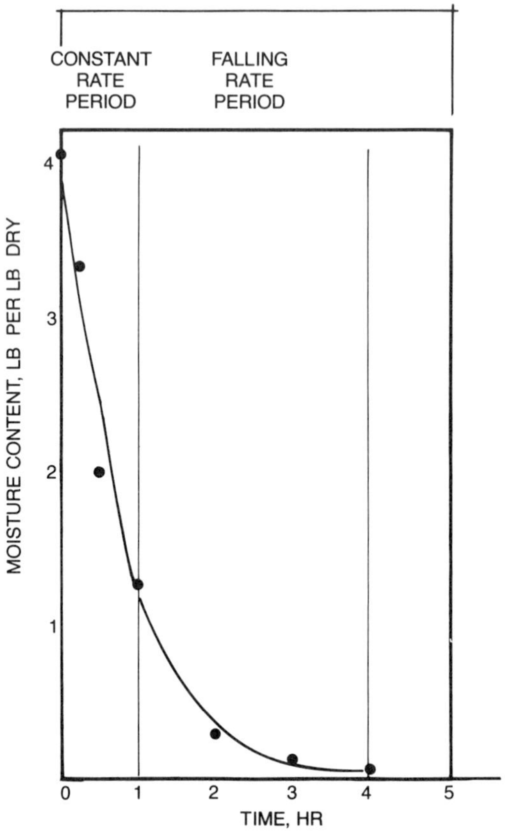

Fig. 9.1. Typical drying rate plot showing constant and falling rate phases.

moisture-transfer term noted above. Diffusion of moisture to the particle surface is one of the most important of these forces (although additional factors are probably involved) and controls the overall drying rate. Other factors affecting drying rate are the shape, size, and arrangement of the material, arrangement with respect to the drying air stream, temperature, relative humidity, and velocity of the air. The critical nature of these forces has led to their expression in various mathematical formulae (Sherwood, 1929; Krischer, 1939) which are reasonably predictive, but which may not apply to all foods under all circumstances.

Spray-Drying

More processed food is dehydrated by spray-drying than by any other drying method. The food to be dried must be a liquid or slurry, which can be fed under pressure, through an atomizer nozzle or onto a rapidly rotating disk atomizer. Ideally, very small droplets of uniform size are formed and passed into a stream of hot (e.g., 300°C—400°F) air. The effect of atomizing is to increase greatly the surface area exposed; thus the rate of drying is increased, accompanied by reduction in various undesirable attributes, such as browning and flavor reversions. The air stream in spray driers may be co- or countercurrent, with a variety of air-flow patterns being possible. Milk products, coffee, and egg products are some of the principal food products dehydrated by this means.

Although the thermal efficiency of spray driers is low, their very rapid drying rates and the fact that they produce dried particles that are hollow, and thus easily rehydratable, recommend this process for many applications.

Drum Drying

The principal application of drum drying in the food industry occurs in the dehydration of potatoes and, to a lesser degree, tomatoes and milk. Drums varying in size from 2 to 10 ft in diameter are used. The surface temperature of these steam-heated drums may reach 280–320°F (Willard and Kluge, 1974), although, due to evaporative cooling, the temperature of the product film during drying seldomly exceeds 200°F. The speed of revolution is approximately 1–5 rpm, with the dried product in contact with the drum for less than one complete revolution; thus the time of drying is usually measured in seconds. The product is distributed over the drum surface by small applicator rolls and is removed subsequently with a doctor blade as a continuous dry sheet, usually containing 8–10% moisture (0.30–0.35 a_w), in the case of dehydrated potato.

Despite the very short drying times, some drum-dried foods may acquire "cooked" flavors and so are suitable only for those applications in which their flavor cannot be detected or is not considered a liability. If such limitations are absent or of little consequence, drum drying is relatively inexpensive and efficient.

Fluidized Bed Driers

Driers of this type have found limited application in the drying of vegetables, especially dehydrated potato granules. In fluidized bed driers, the moist product is introduced continuously onto a grate or porous plate through which flows hot, dry air. As the product slowly moves over this plate in a continuous "bed," it is gently suspended by the hot air stream which facilitates drying.

Cabinet Driers

In these driers, the product to be dehydrated is placed in shallow pans or screens, which are then stacked in a cabinet-type drier. Air is heated and introduced at one end of the unit, and blown across the product at a rate of 7–15 ft/second. Following dehydration, pans or screens are removed from the cabinet and emptied of their dried contents. Certain egg products, fruits, and vegetables are dried commercially in this manner. The principal disadvantage of cabinet driers is that they are batch operations and hence, labor intensive and operation costly.

Belt and Tunnel Driers

Belt and tunnel driers are units similar in design, but differing principally in the means of presenting product to a stream of hot, dry air. In belt driers, the wet product is distributed evenly to a thickness of 1–6 inches on a porous belt through which air circulates in an upward or downward flow pattern. As the product continuously moves through the drier, moisture is removed. Many modern belt driers provide for an upward air flow during an initial drying period and downward air flow during the final drying stage. The purpose of this design is to prevent excessive scattering of the dried product as it leaves the drier. Multistage driers are also useful in situations in which drying conditions must be altered as the food moisture content is reduced (McCabe and Smith, 1967).

Tunnel driers usually provide for horizontal air flow along the length of the drier. Hot, dry air enters the drying chamber at the product discharge end of the tunnel and exits near the point of product entrance. The moist food normally proceeds through tunnel driers on trays supported or stacked on carts moved automatically and continuously along tracks inside the drier. Carts and trays must be introduced into the drier manually, and so, like cabinet drying, tunnel drying is costly.

Foam Drying

The creation of foams for either spray or "mat" drying is based on the advantage of increasing drying rates at relatively low temperatures. This increase is accomplished largely by the extensive enlargement in food surface areas that occurs during foaming. The nature of this process usually requires that it be restricted to liquids.

Thus far, commercial application of foam drying has been limited to fruit juices. If the food to be dried does not form a foam readily, a surface active agent such as an edible monoglyceride, a soluble protein, or an edible gum must be added before a foam of sufficiently stiff structure and stability is obtained. Preconcentration may also help to create stable foam. In the foam mat process, the product plus foaming agent (if necessary) are beaten rapidly to produce the foam, which is then spread onto a porous belt; the latter passes through a tunnel drier. Foam mats are usually 2–3 mm in thickness with drying typically completed with 10–20 minutes at approximately 135°F. This process usually will result in a foam with a density of 0.2–0.3 gm/cc. In foam spray drying, the foam is dried in a conventional spray drier described elsewhere in this chapter.

Rapid drying and superior rehydration are advantages of this type of drying, while the disadvantages are restrictions to the drying of liquids and the lack of efficient heat transfer as a result of the insulating properties of the foam-entrained air.

Freeze-Drying

Over the past 15 years, there have been intensive investigations throughout the world on the application of freeze-drying techniques to commercial food operations. The principal advantages of this process are the elimination of product shrinkage, improved flavor and color retention, and superior rehydratability. The main disadvantage is high cost, as compared to other techniques. From 3 to 8 times as much energy is required to remove a given amount of water. This drawback tends to restrict freeze-drying to relatively expensive foods.

In the freeze-drying process (King, 1973), water is sublimed directly from ice crystals in the frozen product. As a result, the ice is not allowed to melt, which, in more conventional forms of drying, results in shrinkage and distortion. An additional advantage of this process is minimal loss of aroma and flavor constituents as a result of the low temperature at which water is removed (Clark and King, 1971). In most freeze driers, trays of frozen food are placed on heated shelves within a vacuum chamber. As a vacuum of 4.7 mm of mercury or less is drawn in the chamber, heat is applied to the shelf to provide the latent heat of sublimation and so to overcome intermolecular binding of the water molecules. Care must be observed at this point to assure that ice within the food does not melt. As sublimation proceeds, the ice front within the food recedes at a progressively slower rate as a result of the insulating properties of the dried portion. The final product moisture level may be approximately 2–8% (0.10–0.25 a_w or less). This a_w level and the extensive surface area exposed by the departed ice may render the porous product highly susceptible to autoxidation, and special precautions such as inert gas or vacuum packaging may have to be employed. The rate-limiting factors affecting freeze-drying are shown in Fig. 9.2.

Various foods have been commercially dehydrated by freeze-drying; brewed coffee, certain fruits, such as bananas and strawberries, shrimp and diced chicken are dried in this manner. In addition, a variety of freeze-dried foods for hikers and campers are produced.

The optimistic predictions for freeze-drying that were commonly made 10–15 years ago have not been realized. The present high costs of energy and

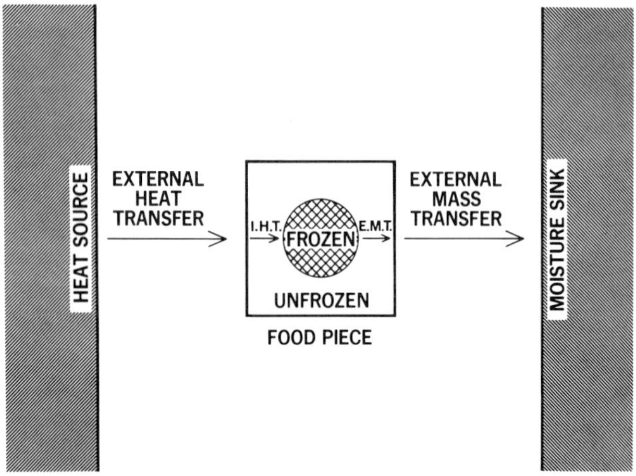

Fig. 9.2. Rate limiting factors in freeze drying. I.H.T., internal heat transfer; E.M.T., external mass transfer (King, 1973).

labor cast some doubt on the wide-scale application of this technique for all but very high-value foods. Despite this, there is hope that a less expensive process will be developed from the intensive research currently being devoted to it. Microwave heating of food to provide energy application directly to the ice front during freeze-drying has proved feasible and effective (Decareau, 1970). Among the other alternatives investigated (Clark and King, 1971) is convective freeze-drying, in which food pieces and a molecular sieve desiccant are placed in a bed in either mixed or layered form. A light gas, such as helium, is then circulated through the bed at moderate velocity and pressure. The gas facilitates transfer of water from the food to the sieve and moves heat from the desiccant to the food. Little or no external heat need be applied to the process, yet the gas can be at a temperature above ambient without melting the water in the food.

Another modification is cyclic freeze-drying (Middlehurst, 1974). A gas of high thermal conductivity, such as nitrogen or preferably helium, is fed into the vacuum chamber for a short time, pumped out again, and the cycle repeated until the food is dried. The rate of sublimation of ice is increased, with a substantial reduction in drying time.

Development of improved rehydration techniques may also enhance the potential of freeze-drying. As noted earlier, many freeze-dried foods rehydrate with greater ease and preservation of original food properties than foods dehydrated by other techniques. Unfortunately, the preservation of the rigid, porous structure that aids rehydration of some foods may hinder rehydration in others. Variations in freezing rate (Longan, 1973) and the composition of the rehydrating solution (Curry *et al.*, 1976) may alleviate this difficulty.

Sun Drying

As noted above, sun drying is probably the oldest dehydration technique known to man and is still practiced in various parts of the world. This procedure requires that drying be sufficiently rapid to reduce a_W to a level at which microbial growth will not occur before drying is complete. Obviously, weather conditions are important, although extended periods of sunlight are not an absolute requirement, since outdoor drying is practiced in arctic climates, where ice may be sublimed directly from the frozen food.

Some products, especially such fruits as apricots, figs, raisins, and dates, continue largely to be sun-dried. Chemicals, for example, sulfur dioxide (SO_2), may be added before drying cut fruits to preserve color, flavor, and certain vitamins and to prevent mold and yeast development. More than 85% of the raisins produced in California are sun-dried on paper trays placed between the vineyard rows (Fig. 9.3). In this case, few attempts are made, or, for that matter, are possible to control the drying rate and final a_W level in the field.

The raisins are normally dried to 17% (a_W = 0.60) or lower moisture, which may require from 2 to 3 weeks, depending on the climactic conditions during the drying period. Gradually, commercial sun drying operations are being replaced by mechanical drying techniques (usually tunnel driers) which afford better process control and improved product uniformity.

Fig. 9.3. Raisins drying on paper mats in vineyard (courtesy of the California Raisin Advisory Board).

Salting

The preservation of food by salting is nearly as old as sun drying. Commerce in salt by ancient Egyptian, Chinese, and Phoenician cultures is well documented (Tannahill, 1973), and wars were waged over the possession of salt deposits during medieval times. Today, salt is one of the least costly food ingredients, as a result of ready availability and mechanized methods of production.

Although economically important at one time, the commercial salting of meats and fish is less practiced today largely as a result of the preservation of these foods by refrigeration. Certain vegetables such as cabbage (sauerkraut) and cucumbers (pickles), which require salting during their processing, continue to be frequently consumed foods in many parts of the world.

Fish may be preserved by salt brining (herring, anchovies) or by "dry salting," in which case the eviscerated fish are exposed to layers of crystalline salt. Both chemical and microbiological deterioration may occur in salted fish. Autoxidation may be a problem in fish containing high levels of fat, such as cod or tuna, and the growth of halophilic bacteria may cause flavor reversions and/or color changes. Deterioration especially may be a problem when sea salt is employed to preserve fish because this material often is heavily contaminated with halophilic and osmotolerant microorganisms.

Brining continues to be an important method of curing hams, although modern preservation methods allow reductions in the amount of salt used. The NaCl content of ham-curing brines normally ranges between 60 and 70% of saturation (0.87–0.82 a_w), according to Deibel and Niven (1958), although other solutes such as glucose, sucrose, and sodium nitrate and nitrate may be present to depress further the a_w level. Brines usually are pumped into the vascular system of the ham and along the bone, followed by total immersion into the brine prior to trimming and smoking. Bacon (Patterson, 1963) is normally processed in a similar manner, although multiple point injection or "stitch pumping" of the curing salts has become common.

As noted above, sauerkraut and cucumber pickles are brined, in the former case by the addition of approximately 2.5% NaCl by weight to the chopped cabbage. In the case of cucumbers, a brine (8% NaCl) is added, plus an additional 9–10 lb of salt per 100 lb of cucumbers (final a_w is approximately 0.87). Higher NaCl concentrations may be used, depending on the type of pickle desired (Frazier, 1967). The purpose of the NaCl in both of these fermentations is to promote the growth of desired, osmotolerant, lactic acid-producing bacteria. The production of lactic acid and the addition of acetic acid (pickles) are the primary vehicles by which these products are preserved. The NaCl added acts primarily as a selective agent to assure that the proper sequence of desirable bacteria develops during the fermentation.

In recent years, salt and acid have been used as preservatives in cucumber pickles, without fermentation.

Other applications of salting occur in the manufacture of many types of cheese, which may be soaked in a brine of 20–25% NaCl during manufacture or rubbed with salt during ripening. Usually, the purpose of this treatment is to limit the growth of unwanted, proteolytic bacteria that possess a higher minimal a_w for growth than the desired lactic acid bacteria.

CONCENTRATION BY WATER REMOVAL

The concentration of foods does not always produce measurable reductions in a_w. With a product such as orange juice, 70–75% of the water content may be removed, but this still leaves sufficient moisture to provide an a_w in excess of 0.99. The principal advantage in concentrating this material is to achieve economically beneficial reductions in volume and weight. Usually, such products are stabilized by other means, such as freezing, because their high a_w level would otherwise encourage deleterious changes.

As pointed out by Karel *et al.* (1975), there are three general techniques for the concentration of foods: evaporation, freeze-concentration, and membrane separation. All, of course, involve water removal.

Evaporation

In this process, heat is applied to a food to boil away a solvent, usually water. Often a vacuum is employed to lower the boiling point if the food is sensitive to high temperatures. Although evaporation is the most common method of concentration, it has the disadvantage that off-flavors may develop, and desirable flavors may be lost as a result of volatilization.

Freeze-Concentration

Freeze-concentration involves water separation by freezing, in which case ice crystals are removed from the food by filtration or centrifugation. This method has been utilized to concentrate coffee and tea preparatory to freeze-drying. Although costly, it involves low temperatures and, thus, flavor changes are avoided.

Membrane Processing

The basis for this process is the use of a semipermeable membrane that differentially allows water to pass through the membrane while retaining food

TABLE 9.1.

a_w of Some Concentrated Foods[a]

Orange juice concentrate	0.92
Maple syrup	0.86–0.83
Canned vegetable soup	0.98
Sweetened condensed milk	0.85–0.89
Evaporated milk	0.97
Tomato paste	0.97

[a] Personal data of the authors.

solids. In most commercial processes of this type, a pressure difference across the membrane provides the driving force and determines, along with membrane resistance, the rate of separation. This method of concentration is not used extensively at present, but wider application is expected as membrane technology expands.

Concentrated foods, such as maple syrup (Table 9.1), may possess a water content not too dissimilar to that of citrus juice concentrates and yet be perfectly stable as a result of low a_w. In this example, water evaporated by boiling, which also develops the characteristic flavor and color of this product, results in a large rise in sugar concentration and a corresponding fall in a_w. Here, the preservation effect of low a_w need not be augmented by other stabilizing factors. The principal difference between these two methods of concentration lies in what remains following water removal. In concentrated orange juice or fruit, or in vegetable purees, the levels of hygroscopic solutes remaining are very low, whereas in the case of maple syrup, concentration of sugars in the product is substantial, and so a relatively low a_w is obtained (about 0.86).

As noted above, one of the major applications of concentration processing is to reduce the moisture content of foods prior to drying. This step often precedes dehydration because concentration normally is a less costly means of water removal than most drying techniques.

A detailed description of the various types of concentrators commercially available is beyond the scope of this book, but excellent reviews of this subject exist in volumes by Perry *et al.* (1963), McCabe and Smith (1967), and Karel *et al.* (1975). It is apparent from these works and other literature that much process development work has been devoted to this type of food processing.

Preservation by Intercomponent Moisture Transfer

Individually dried, multicomponent food systems usually destined for "complete meal" consumption have become increasingly popular in recent years. Convenience and economics in shipment and manufacture plus the inherent stability of these foods recommend them. In simple, two-component systems containing ingredients dissimilar in a_w, the eventual a_w levels in the components (in a vapor-proof package) are predictable (see below). The rate at which equilibration is established will depend on the moisture levels and other properties of the mixture. For example, moisture may migrate from a dried diced vegetable to lower a_w components very slowly because of case hardening or a high degree of water binding. Conversely, dried fruit possessing a relatively high a_w, when mixed with crystalline sucrose or a concentrated sugar solution, may equilibrate rapidly. Systems containing ingredients with extensive surface areas will also tend to equilibrate at a higher rate. The rate of equilibration must be such that deleterious changes such as browning, enzymatic changes, and microbial growth are prevented during the moisture-adjustment phase.

Grover (1947), in his studies with various confections, developed formulas for predicting the final a_w of foods containing several components. Unfortunately, this work was limited to solutions of known chemical composition and vapor pressure. Somewhat later, Salwin and Slawson (1959) described a procedure in which sorption isotherm slopes of individual components are employed to calculate the equilibrium moisture distribution in combinations. Such data may be critical in assuring the stability of packaged products during storage. The equation derived for calculating the equilibrium relative humidity (E.R.H.) of a mixture (R_m) is

$$R_m = \frac{R_A S_A W_A + R_B S_B W_B}{S_A W_A + S_B W_B} \qquad (2)$$

The initial E.R.H. values of components A and B are described as R_A and R_B; W_A and W_B represent the dry weights of A and B, and S_A and S_B are the slopes of the sorption isotherms. If more than two components are involved, additions to the denominator and numerator of the above equation are made.

Salwin (1962) employed Eq. (2) to predict the moisture distribution in a dehydrated chicken stew mix. These data (Table 9.2) show the initial, predicted, and observed relative humidity and moisture levels in various components of the stew. The close agreement obtained between predicted and observed values indicates that Eq. (2) has practical value in estimating relative humidity, and thereby, the substantial lack of interactions between the mixture components. As pointed out by Salwin (1962), the E.R.H., which provides the greatest stability to the most sensitive ingredient, should be used as a guide in formulating to a specific and desired value for the finished mixture.

In multiple component mixtures, storage conditions that minimize stability problems, such as browning, lipid oxidation, flavor deterioration, etc., are usually in harmony with the final use of the mixed product. Occasionally, however, rehydration of a component may be adversely affected by equilibration of the

TABLE 9.2.

Predicted and Observed Moisture Transfer in a Dehydrated Chicken Stew Mixture at 72°F[a]

		Relative humidity (%)		Moisture (% of solids)		
					Final	
Ingredient	Weight (grams)	Initial	Final predicted	Initial	Predicted	Observed
Chicken	39.9	1.1	7.6	1.07	3.17	3.18
Potatoes (diced)	30.1	28.4	7.6	5.88	3.45	3.64
Lima beans	15.2	7.7	7.6	3.68	3.68	3.90
Cream sauce base	5.0	18.8	7.6	3.49	2.60	2.54
Chicken soup and gravy base	4.0	37.0	7.6	2.49	1.17	1.19
Mixture, calculated total water (grams)				2.90	2.94	3.03

[a] Salwin, 1962.

ingredient to a specific a_w or E.R.H. For example, a hydrated soup mixture may contain a component, dried green pepper, which gives a tough or leathery sensation when tasted. These situations can usually be avoided by altering the process that the component receives prior to drying. Alternatively, the sensitive ingredient may be packed in a separate, vapor-proof package to prevent a_w equilibration with other components in the mixture.

INTERMEDIATE MOISTURE FOODS

Definition

Various definitions of intermediate moisture foods (IMF) have been suggested since this term entered the food scientists' lexicon. Possibly the best definitions rely minimally on a_w or moisture content and emphasize operational parameters. IMF (for either human or animal consumption) are sufficiently plastic to eat without further hydration and of sufficiently low a_w to prevent bacterial growth. These foods may, however, be susceptible to mold growth, enzymatic degradation, or nonenzymatic browning unless appropriate preventive measures are taken. Intermediate moisture foods normally range in a_w from 0.7 to 0.9 and in water content from 20 to 50% (Karel, 1973). A list of several commonly available IMF is given in Table 9.3. Many of the IMF listed in this table have been produced and consumed for centuries, whereas others have been introduced more recently as the technology for their fabrication has been developed. Usually, the

TABLE 9.3.

Examples of Intermediate Moisture Foods

Prunes
Dried apricots
Jams and jellies
Semimoist pet foods
Fruitcake
Pemmican
Dry sausages (e.g., pepperoni)
Marshmallows
Country-style ham

latter involve the addition of a humectant, and we refer to these types of foods as fabricated IMF.

Principles of IMF and Their Application

Although a_w concepts need not form an essential part of an IMF definition, factors relating to a_w are of central importance in understanding the principles that underlie IMF technology. The aim of this technology is to reduce the a_w of a food to a range in which most bacteria in foods will no longer grow (Plitman *et al.*, 1973). As pointed out in Chapter 5, bacteria, other than halophiles, will not grow at 0.83 a_w or below, and most are inhibited markedly at 0.90 a_w or less. Foods supporting the growth of halophilic bacteria at a_w levels below 0.83 would normally be considered too salty for use. Because many molds proliferate at a_w levels $\leq$0.83, antimycotic agents such as sorbic or benzoic acids are often added to IMF. Alternatively, a_w may be lowered by a humectant possessing some degree of antimycotic activity, such as 1,3-butanediol or propylene glycol.

As noted earlier in this chapter, other deleterious changes may occur in IMF and appropriate steps must be taken to prevent them. Lipid oxidation is normally controlled by antioxidants or by inert gas packaging (Labuza, 1972). A mild heat treatment will enhance stability by preventing enzymatic degradation. Especially important is the enclosure of IMF in moisture-proof packages to prevent interchange of water between the food and the atmosphere. The use of glycerol or other liquid humectants also may provide some protection from nonenzymatic browning.

Several solutes serve as safe and effective agents for reducing a_w levels in IMF. Sodium chloride and sucrose have been the most widely used for this purpose. On a mole-for-mole basis, NaCl is the more effective, but sucrose is organoleptically more compatible with fruits, and this has been historically its major application. In addition, sucrose now plays a major role in the develop-

TABLE 9.4.

Typical Composition of a Semimoist, Intermediate Moisture Pet Food[a]

Ingredient	Parts by weight
Chopped meat by-products	32.0
Defatted soy flakes	31.0
Sucrose	21.7
Flaked soy bean hulls	3.0
Dicalcium phosphate	3.0
Dried nonfat milk solids	2.5
Propylene glycol	2.0
Tallow	1.0
Mono- and diglycerides	1.0
Sodium chloride	1.0
Potassium sorbate	0.3
Dye	0.006
Garlic	0.20
Vitamin and mineral premix	0.06

[a] Potter, 1970.

ment of fabricated, semimoist pet foods. In a "typical" IMF as shown in Table 9.4, sucrose is the primary solute, although small amounts of glycerol, propylene glycol, and NaCl also may be added. Sodium chloride is used mainly to reduce a_w in meat products for human consumption, such as salt-cured hams, bacon, and some types of sausages.

Recent studies of intermediate moisture meat prepared by high temperature equilibration with solutions containing NaCl and glycerol, so that their a_w levels are poised between 0.82 and 0.86, have shown that both cross-linking and breakdown of proteins occur during storage (Obanu *et al.*, 1975a,b) at 38°C. The development of soluble hydroxyproline as a result of collagen breakdown corresponded with an increase in tenderness during subsequent storage. Although these results indicate that textural changes will not affect the acceptability of IMF meat, the protein alterations observed could result in nutritional changes that might be unacceptable.

The very nature of IMF products is based on increasing their shelf life, and thus microbiological stability has been emphasized, sometimes to the exclusion of other potentially deleterious effects such as color changes, lipid oxidation, and, as noted above, nutritional losses. Because these materials have been marketed as pet foods, these potentially negative effects have perhaps been thought to be of secondary importance. However, if such products are to be considered seriously as alternatives for large-scale human consumption, many of these potential negative attributes will have to be considered and possibly corrected. The effect of lowered a_w levels on the nutritional quality of foods is discussed further in Chapter 4.

Glycerol and several other polyhydric alcohols have also received consideration as humectants in IMF. Unfortunately, many of these compounds produce a significant depression of food a_w only at concentrations above their flavor threshold, an outcome that restricts their usefulness in many foods. To overcome this difficulty, combinations of solutes may be employed. Usually, mixtures of NaCl, glycerol, and/or sucrose are used, although propylene glycol may also be substituted for one or more of these ingredients. The water-binding properties of the natural food components may be altered to obtain some reduction in a_w. Methods of achieving this are relatively undeveloped at present, but chemical or physical modification of proteins, for example, by urea treatment or heating at relatively low pH levels, is worthy of consideration.

Various methods of introducing or infusing humectants into IMF have been discussed by Kaplow (1970) and Brockmann (1970). Basically, these procedures involve an initial drying step followed by soaking the food in an aqueous solution of the humectant. The food is subsequently drained and packaged. In terms of the sorption isotherm, an initial and limited hydration is involved. An alternative method of achieving solute infusion is by a desorption process in which a moist food is equilibrated in a solution of lower a_w. This process may be facilitated by heating or cooking the food in the solution and, as pointed out by Kaplow (1970), is based on the assumption that the final concentrations of solute in the food and in the infusing solution will be similar. Usually, the cooked food is refrigerated during at least part of the infusion process. The final a_w of the IMF does not appear to depend on the method or order of humectant mixing (Sloan *et al.*, 1976).

Labuza (1972) and Acott and Labuza (1975) have emphasized the critical relevance of the hysteresis loop in the manufacture of IMF. For example, a dried food (absorptive mode) that is being treated to obtain a specific a_w level may possess the same solids and a_w but a lower moisture content than a similar nondried food that is being equilibrated with a reduced a_w solution (desorption mode). Of the two methods, the latter probably offers greater advantages because it is less costly and enzymes are inactivated if cooking is included in the process. Also, rapid equilibration is not dependent on the physical characteristics of a porous, dried food, which could be refractory to partial rehydration by a sometimes viscous humectant solution. Unfortunately, work on intermediate moisture foods has shown that microbial growth also may be more rapid in foods prepared by water desorption, even when parity of a_w levels occurs.

An interesting variation on the desorption infusion procedure has been proposed by Li *et al.* (1974). In this process, peaches are immersed in a 30% sucrose syrup containing $NaHSO_3$ and potassium sorbate as preservatives. A vacuum of 27 inches is drawn to remove tissue gases, which are promptly replaced by the syrup during a subsequent release of the vacuum. Further reduction in a_w and total moisture is achieved by dehydration in a tray drier to 30–40% moisture. Although there were varietal differences, a satisfactory product resulted, with a shelf life of up to 9 months. To our knowledge, this process has not been commercially exploited.

Outlook

Despite impressive inroads in the commercial pet food market in Europe and the United States, the commercial fabrication of IMF for human consumption, using currently approved humectants, is viewed pessimistically by the authors. Certain products, such as pemmicanlike, fabricated foods, have entered (and often promptly left) the market. Certainly, there seems to be little consumer interest in these foods. In our view, the principal reason for this discouraging picture is the lack of humectants that are safe, cheap, effective, and without flavor, color, or nutritional drawbacks. When and if such compounds become available, fabricated IMF could become a commercial and economically feasible entity. Until such time, we must conclude that, while technically interesting, fabricated IMF for human consumption is not a practical reality. On the other hand, the traditional nonfabricated foods, such as those noted earlier in this section continue, justifiably, to be acceptable products.

REFERENCES

Acott, K. M., and Labuza, T. P. (1975). Microbial growth response to water sorption preparation. *J. Food Technol.* **10,** 603–611.

Brockmann, M. C. (1970). Development of intermediate moisture foods for military use. *Food Technol. (Chicago)* **24,** 896–900.

Clark, J. P., and King, C. J. (1971). Convective freeze drying in mixed or layered beds, *Chem. Eng. Prog.* **67,** 102–111.

Curry, J. C., Burns, E. E., and Heidelbaugh, N. D. (1976). Effect of sodium chloride on rehydration of freeze-dried carrots. *J. Food Sci.* **41,** 176–179.

Decareau, R. V. (1970). Microwave energy in food processing operations. *Crit. Rev. Food Technol.* **1,** 199–224.

Deibel, R. H., and Niven, C. F., Jr. (1958). Microbiology of meat curing. I. The occurrence and significance of a motile microorganism of the genus *Lactobacillus* in ham curing brines. *Appl. Microbiol.* **6,** 323–327.

Frazier, W. C. (1967). "Food Microbiology." McGraw-Hill, New York.

Grover, D. W. (1947). The keeping properties of confectionery as influenced by its water vapor pressure. *J. Soc. Chem. Ind., London* **66,** 201–205.

Kaplow, M. (1970). Commercial development of intermediate moisture foods. *Food Technol. (Chicago)* **24,** 889–893.

Karel, M. (1973). Recent research and development in the field of low moisture and intermediate moisture foods. *Crit. Rev. Food Technol.* **3,** 329–373.

Karel, M., Fennema, O. R., and Lund, D. B. (1975). "Principles of Food Science. Part II. Physical Principles of Food Preservation." Dekker, New York.

Keey, R. B. (1972). "Drying: Principles and Practice." Pergamon, Oxford.

King, C. J. (1973). Freeze drying in food dehydration. *In* "Food Dehydration" (W. B. Van Arsdale, M. V. Copley, and A. I. Morgan, Jr., eds.), pp. 161–200. Avi Publ. Co., Westport, Connecticut.

Krischer, O. (1939). Physical problems in the drying of solid porous materials. *Chem. Appl.* **26,** 17–23.

Labuza, T. P. (1972). Stability of intermediate moisture foods. I. Lipid oxidation. *J. Food Sci.* **37,** 154–159.

Li, K. C., Heaton, E. K., Boggess, T. S., Jr., and Shewfelt, A. L. (1974). Improving fruit quality with intermediate moisture process. *Food Prod. Dev.* **8,** 52–54.

Longan, B. J. (1973). Effect of processing variables on quality of freeze-dried carrots. Ph.D. Thesis, Texas A&M University, College Station, Texas.

McCabe, W. L., and Smith, J. C. (1967). "Unit Operations of Chemical Engineering." McGraw-Hill, New York.

Middlehurst, J. (1974). Freeze drying. I. The CSIRO process for cyclic freeze drying. *CSIRO Food Res. Q.* **34,** 31–34.

Obanu, Z. A., Ledward, D. A., and Lawrie, R. A. (1975a). The protein of intermediate moisture meat stored at tropical temperature. I. Changes in solubility and electrophoretic pattern. *J. Food Technol.* **10,** 657–666.

Obanu, Z. A., Ledward, D. A., and Lawrie, R. A. (1975b). The protein of intermediate moisture meat stored at tropical temperature. II. Effect of protein changes on some aspects of meat quality. *J. Food Technol.* **10,** 667–674.

Patterson, J. F. (1963). Salt tolerance and nitrate reduction by micrococci from fresh pork, curing pickles and bacon. *J. Appl. Bacteriol.* **26,** 80–85.

Perry, R. H., Chilton, C. G., and Kirkpatrick, S. D., eds. (1963). "Chemical Engineers Handbook," 4th ed. McGraw-Hill, New York.

Plitman, M., Park, Y., Gomez, R., and Sinskey, A. J. (1973). Viability of *Staphylococcus aureus* in intermediate moisture foods. *J. Food Sci.* **38,** 1004–1008.

Potter, N. N. (1970). Intermediate moisture foods: Principles and technology. *Food Prod. Dev.* **4,** 38, 41, 44, 45, and 48.

Salwin, H. (1962). Moisture in deteriorative reactions of dehydrated foods. *Freeze-Drying Foods, Proc. Conf., 1961* pp. 58–74.

Salwin, H., and Slawson, H. (1959). Moisture transfer in combinations of dehydrated foods. *Food Technol. (Chicago)* **13,** 715–718.

Sherwood, T. K. (1929). The drying of solids. *Ind. Eng. Chem.* **21,** 112–116.

Sloan, A. E., Waletzko, P. T., and Labuza, T. P. (1976). Effect of order-of-mixing on a_w lowering ability of humectants. *J. Food Sci.* **41,** 536–540.

Tannahill, R. (1973). "Food in History." Stein & Day, New York.

Van Arsdale, W. B., Copley, M. V., and Morgan, A. I., Jr., eds. (1973). "Food Dehydration," Vol. 1. Avi Publ. Co., Westport, Connecticut.

Willard, M., and Kluge, G. (1974). Potato flakes. *In* "Potato Processing" (W. F. Talburt and O. Smith, eds.), pp. 463–512. Avi Publ. Co., Westport, Connecticut.

10

Packaging, Storage, and Transport

The storage and transport of foods and commodities demand the maintenance of conditions under which deterioration from whatever cause is kept to a minimum. The desirable conditions have been outlined in earlier chapters and some of the problems that arise in maintaining these conditions will now be discussed. These problems are influenced profoundly by the temperature and humidity of storage and by the type of container or package in which the foods are stored.

BULK STORAGE OF COMMODITIES

Commodities stored in bulk are predominantly "dry." For grains and other seeds, this may mean an a_w as low as 0.625 (see Fig. 10.1), a level at which they will be protected from mold growth (Poisson and Guilbot, 1963) but, in most cases, remaining susceptible to insect attack.

An initially uniform a_w throughout a stored product will remain so only as long as its bulk is isothermal. If the temperature of the enclosure differs from that of the bulk stored product, as will inevitably happen in the absence of expensive and extensive precautions, temperature gradients will be formed. These will result in the migration of water vapor from the warmer to the cooler region, so that a_w may rise substantially in the latter area to the point where molds can grow and, in some cases, to where condensation of liquid water will occur. A further possible consequence is travel of warm, moist air by convection into a cool installation, contributing additional water to the system.

Temperature gradients result from diurnal or seasonal temperature variations. Where diurnal variations in temperature result in moisture migration, the

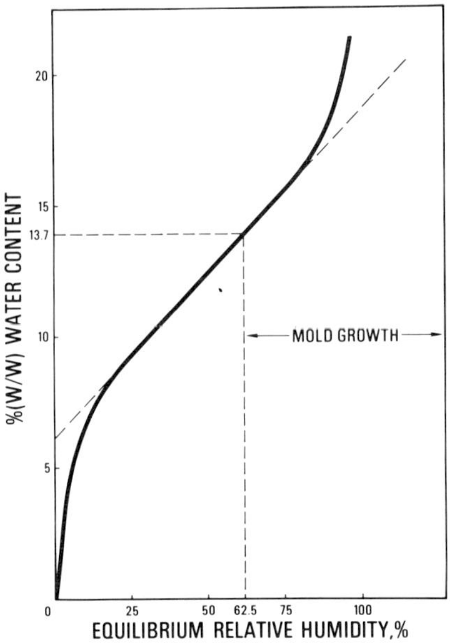

Fig. 10.1. Sorption isotherm of wheat at 25°C. Minimal levels for mold growth are marked by broken lines (Poisson and Guilbot, 1963).

gradients in product a_w may often be reduced to acceptable levels by aerating the installation.

In temperate zones, moisture migration in silos is a common problem, since grain, stored after harvest, is cooled from outside at the onset of winter. Zones of high a_w develop towards the top and sides of the silos and become foci for mold growth. These zones may be conveniently dissipated by drawing cool ambient air through the stack so that isothermal conditions are restored.

In tropical and subtropical regions, high average relative humidities make the drying of grains and seeds prior to storage particularly difficult. Increasing use is made of drying with warm air, supplied by oil burners, to reduce the a_w of the commodity in or prior to storage to below 0.70 a_w. Even properly dried commodities will remain sound in such climates only if adequate barriers against ingress of water and water vapor are provided. The possibility of moisture migration within the storage unit exists, and one of the most economical ways to avoid this through isothermal storage is to use underground storage facilities.

Insects are a major cause of loss during storage because many species develop at levels of a_w below 0.30, which is far below that inhibiting mold growth. Infestation under such conditions can lead to the metabolic production of water by the insects and, hence, local areas of elevated a_w, where mold growth now becomes possible. A cycle of heat and moisture production by both insects and fungi can thereby result, leading to extensive regions of damage. Because insects can influence the a_w level in commodities, brief mention will be made of methods for their control.

Insect infestation of bulk stored commodities is controlled by low temperature storage, by anaerobiosis, and by chemicals. Lowering the temperature to < 15°C effectively prevents the multiplication of insects which at temperatures of 25°C develop rapidly at the normal a_w of stored grain. In many temperate regions, temperature reductions to such levels may be obtained readily by ventilation with cool night air.

When cooling by aeration, the temperature reached by the commodity will depend upon the relative humidity of the air as well as on its temperature. Bulk grain behaves like a wet-bulb thermometer in its response to temperature and humidity and the cooling capability of air for grain aeration is best indicated by its wet-bulb temperature (Griffiths, 1967). Cooling by ventilation will normally result in some loss of water from the commodity, a loss which may make a useful contribution to the drying process.

When the climate is such that the required grain temperature cannot readily be achieved by simple aeration, the use of mechanical refrigeration to cool the airstream may be economically feasible. As well as suppressing the development of insects, it will retard the spoilage by mold of high-moisture grain that must be stored prior to drying.

All insects require atmospheric oxygen for normal development and anaerobic storage thus provides a means of controlling them. An oxygen concentration of less than 2% is effective against all insect species encountered in commodity storage. The death rate may be low under these conditions, and there is no residual effect, so anaerobiosis must be maintained throughout storage. In principle, these conditions are attained most readily by making the storage facility gastight and flushing it with nitrogen. In practice, however, it has been found very difficult to make existing silos gastight, and so the nitrogen must be fed continuously into the silo. Anaerobic storage seems likely to increase in popularity as the use of chemical fumigants and insecticides meets with growing official disapproval on world markets.

TRANSPORT

Condensation resulting from moisture migration has proved a problem in the successful transportation of some foods and commodities through differing climatic zones. For example, in a shipment of pea beans from Detroit to London, cooling during passage through the Labrador current zone was followed by increases in atmospheric temperature and humidity in the Gulf Stream zone (Thompson *et al.*, 1962). Condensation on the inside surface of the hold occurred, and the water dripped onto the bags of beans. The relative humidity of the air above the bags rose above 90%, but was reduced to a safe level by aeration of the hold for 8–11 hours during the middle of each day.

Similar condensation problems have been encountered in ISO* containers of canned foods (Middlehurst, 1968). Here, it is not the food that deteriorates but the can and label, by corrosion and mold attack, respectively. It is not usually practicable to reduce the total water load on a container after loading. Therefore, prevention of condensation depends primarily upon drying the container before loading, restricting the water content of cartons and pallets, and avoiding extremes of ambient temperature. The latter is achieved by avoiding land transport on very cold, clear nights, and, on shipboard, by stowing containers below deck and away from heated or refrigerated areas.

Relatively small decreases in air temperature may be required to cause condensation in an atmosphere that has previously been in equilibrium with a commodity whose a_w level was low enough to prevent mold growth. The psychrometric chart in Fig. 2.7 (Chapter 2) shows that air at 70% E.R.H. and 30°C becomes saturated and, hence, prone to moisture condensation if its temperature falls by only 6°C.

REFRIGERATED STORAGE OF BULK PRODUCTS

Most foods stored at chill temperatures, i.e., close to but above their freezing points, depend predominantly on low temperatures for their commercial shelf life. However, in those which are not protected by water vapor-impermeable packaging, the relative humidity of the storage atmosphere can be critical in assuring maximal shelf life.

Where the object of chilling is to retard microbial growth, as is the case with carcass meat, a relative humidity that leads to a reduction in a_w level provides a further restraint, but may result in changes in appearance and increased weight loss that may not be acceptable. Evaporative cooling is important in the chilling of meat carcasses postslaughter. The drying that results is not a function of the relative humidity alone, but is markedly influenced by the velocity of the circulating refrigerated air, increasing, but not proportionately, with increasing air speed.

During subsequent chilled storage at close to −1°C, control of both relative humidity and air speed is essential to prevent excessive dehydration, on the one hand, or accelerated microbial growth, on the other. At this temperature, growth of the important spoilage bacteria is severely restricted when the water content of surface tissue is held below 90% dry weight (Scott, 1936). This corresponds to

*International Organization for Standardization.

TABLE 10.1.

Storage of Fruits and Vegetables: Ideal Conditions of Temperature and Relative Humidity[a]

		Relative humidity		
Group	Storage temperature (°C)	A (65–70%)	B (85–90%)	C (95+%)
1	−1 to 0		Apples (non-chilling-sensitive varieties), nuts	Apricots, grapes, well-matured plums, peaches, figs
2	−1 to 0		Pears[b]	
3	0	Onions and garlic (after curing)	Beans (lima), mushrooms, loquats	Artichokes, asparagus (short term), red beets, all berries, broccoli, cabbages, carrots, cauliflower, celery, cherries, sweet corn, leafy greens, horseradish, kohlrabi, green leeks, lettuce, parsley, parsnips, green peas, radish, rhubarb, silver beet, turnips[b]
4	3		Apples (chilling-sensitive varieties)	Asparagus (long term)
5	5		Oranges, mandarins, potatoes (following curing at 15–20°C)[b]	
6	5		Rock melons, watermelons, persimmons, guavas	
7	7		Avocados, honeydew melons, olives, firm ripe tomatoes, papaws	French beans, cucumbers, egg fruit, pepper (sweet) capsicums
8	12	Marrows, pumpkins, mature squash	Grapefruit, lemons, limes, sweet potatoes[b]	
9	12		Mature green tomatoes, mangoes, pineapples	

[a] Compiled from Hall (1973) and Mitchell *et al.* (1972).

[b] Groups that should not be exposed to ethylene concentrations >0.01 μl/liter (W. B. McGlasson, personal communication).

about 0.96 a_w. Additional control of microbial growth and, hence, longer storage life may be obtained by increasing the carbon dioxide concentration in the air to 10–20%.

For chilled storage of living tissues, high R.H. values allow retention of turgor and crispness, and microbial spoilage rates are not usually affected unless or until the tissues have been injured or have broken down. The temperature and relative humidity requirements for the storage of a range of fruits and vegetables are summarized in Table 10.1. For most, the optimal relative humidity for storage is in the range 90–95%. Exceptions include melons (85–90%) and garlic and onions (65–70%). Optimum temperatures for storage vary greatly, from close to the freezing point for the majority to 10°C or higher for many tropical fruit. To obtain maximal shelf life, fully controlled atmosphere storage may be applied, in which not only temperature and relative humidity are optimized, but also oxygen, carbon dioxide, and ethylene concentrations.

The most important food to be presented in the frozen state without packaging is fish, whole or gutted. Both dehydration and oxidation proceed relatively rapidly at −30°C, unless the fish is protected by glazing. The frozen fish is glazed by dipping or spraying several times in water to form a covering of ice which, although it sublimes slowly in frozen storage, effectively restricts both loss of moisture from the fish and uptake of oxygen from the atmosphere.

PACKAGED FOODS

Many of the problems referred to above in relation to the quality of bulk or non-packaged foods can be avoided by appropriate packaging, which means, in many cases, that the size of individual units will be much smaller. What is appropriate will depend upon the properties of the food, the changes packaging is intended to prevent, and the distribution environment. Of particular concern in the present context are changes that are influenced by the a_w of the packaged food, and these include modifications to appearance, taste, odor, and texture resulting from both microbiological and chemical action.

Food packages were once expected simply to contain and identify the contents and provide them with an adequate degree of protection. With advances in packaging technology, however, packages are now, in addition, often expected to provide nutritional and content information and literally to sell the product. These functions have assumed such importance that packaging considerations become an integral part of the development of many new products. Indeed, a new food product may be formulated expressly to take advantage of a packaging innovation. For some foods, packaging films have come to assume an active role. Products may now be conditioned within their packages by permitting the selective passage of certain gases through the packaging film to produce desired product characteristics. In effect, packaging becomes a processing step.

Although maintenance of a desired a_w is the main present concern, there are other specialized requirements that may have to be considered in choosing packaging materials and systems. These include protection from light or oxygen; retention of preservatives; the fragility of the product; and the ease with which the package can be filled, handled, and stored. These and other functional requirements of packages are discussed by Bulmer (1965), Davis (1970), and Karel (1973).

As correctly sealed cans and bottles give virtually complete protection to the contents against the ingress of gases, moisture, and microorganisms, attention will be restricted to flexible packaging materials that exhibit a very wide range of vapor and gas permeabilities.

MEASUREMENT OF WATER VAPOR PERMEABILITY

Of the various techniques available for the measurement of permeability of flexible films, the most frequently used are gravimetric. In these tests, a circular sample of the film is sealed with wax across the opening of a dish, usually of aluminum, containing a desiccant. The assembly is stored under conditions of controlled temperature and relative humidity and the moisture uptake followed by weighing at intervals. While simple and relatively inexpensive, such methods are usually slow. Other procedures for measuring vapor permeability include the use of radioactive water, electrolytic hygrometers, electrolytic moisture meters, infrared absorption, thermal conductivity, and gas chromatography.

The range of conditions under which flexible film packaging is used is very wide, and it becomes necessary to determine moisture permeability, not only at high and low relative humidities, but also at temperatures from above 20°C to −20°C and below. While such tests provide useful data on flat samples of packaging material, the actual permeability when the films are used under commercial conditions may be very different. Folds or creases, heat seals, the application of less permeable overlabels, and many other factors may drastically alter predicted performance. Additionally, some packaging materials are susceptible to damage during package fabrication and handling, and this usually results in higher water vapor transmission rates (W.V.T.R.). There are standard methods for obtaining creased samples of packaging materials that are used to predict package performance, but it is usually advisable to measure transmission rates on made-up packages. The W.V.T.R. of different packages may be compared by measuring the weight losses or gains of packaged products stored under controlled conditions. Alternatively, they may be measured directly by following weight changes in packages filled with desiccant. If desired, the atmosphere within a package or the change in a_w of a packaged food may be measured over a period of time by including small hygrometric sensors or moisture-sensitive chemical indicator devices in the package (see Chapter 2).

Water vapor permeability data are reported in terms of the permeability constant, P (the weight of water diffusing through the material as a function of thickness, area, time and vapor pressure difference across the material) or as W.V.T.R. The latter is expressed as weight of water diffusing through unit area per unit time and the relative humidity and material thickness must be specified separately. Temperature must be specified in each case.

The relationship implicit in the permeability constant formula does not hold for coated or laminated films or where the material and the vapor interact, as occurs between hydrophilic films and water vapor. Consequently, W.V.T.R. is more commonly used in characterizing packaging materials, and it is necessary to choose test temperatures and humidities that will be most appropriate to the particular applications.

To deal with this situation, in which the transfer rate of water through a film is influenced by changes in the a_w of the food and the relative humidity of the air, Karel (1973) developed the following formula to predict the moisture content of the stored product:

$$a_w = f\,(a_{w_0}, t, \text{R.H.}, k_1 \ldots k_n, T \ldots)$$

where a_{w_0} = initial a_w; T = temperature (°K); R.H. = relative humidity; $k_1 \ldots k_n$ = sorptive and diffusion constants; t = time (hours). Good correlations were obtained between predicted and experimental values.

PERMEABILITY OF PACKAGING MATERIALS

The basic flexible materials of importance in packaging food are aluminum, plastics, regenerated cellulose, and paper. Aluminum foil in thicknesses below about 0.001 inch contains pores that permit vapor diffusion; the foil itself is impermeable in the absence of pores. Most aluminum foil used in laminates is 0.00035 inch thick, and thus contains large numbers of these pinholes. However, since these pores are covered by layers of other materials on each side of the foil, the overall permeability of the laminate is near zero. Films based on plastics and on regenerated cellulose vary widely in permeability to water vapor and other gases. Examples of water vapor permeability of a selection of films are given in Table 10.2. The range of permeabilities exceeds 70,000-fold.

Table 10.3 provides examples of both water vapor and oxygen permeabilities for a smaller range of films. In this table, which does not include the very permeable Cellophan PT600, the range of water vapor permeabilities is about 30,000-fold, while oxygen permeability varies some 1000-fold between films.

TABLE 10.2.

Water Vapor Permeability of Various Films at 20–25°C[a,b]

Material	P in $\frac{\text{gm}\cdot\text{cm}}{\text{cm}^2\cdot\text{s}\cdot\text{bar}}$
Polychlorotrifluoroethylene	1.04×10^{-11}
Polyvinylidene chloride	1.25×10^{-11}
Polypropylene	1.46×10^{-10}
Polytetrafluoroethylene	1.87×10^{-10}
Polyvinyl fluoride	1.71×10^{-9}
Polyvinyl chloride	2.74×10^{-9}
Polystyrene	5.62×10^{-9}
Polyacetal	5.83×10^{-9}
Polycarbonate	8.33×10^{-9}
Cellulose acetate	7.01×10^{-8}
Cellophan PT600	7.78×10^{-7}

[a] Niebergall, 1974.
[b] Film thickness = 40 μm. Relative humidity differences = 85 to 0%.

TABLE 10.3.

Oxygen and Water Vapor Permeability of Some Packaging Films[a]

	Permeability [cc(STP)/m²/24 hours/cm Hg at 25°C, 65% R.H.]	
Film material	Oxygen	Water vapor[b]
Low-density polyethylene (0.001-inch)	120	3200
Polypropylene (0.001-inch)	52	1500
Polyvinyl chloride (0.001-inch)	39	5300
Nylon 11 (0.001-inch)	4.6	4700
Polyethylene terephthalate (0.001-inch)	1.1	4400
Polyvinylidene chloride (0.001-inch)	0.11	160
400 MSAT cellulose	1.3	1200
400 MXXTA cellulose	0.092	240

[a] Davis, 1970.
[b] The amounts of water vapor permeating film materials are usually expressed in grams. cc(STP) water vapor $\times 8.032 \times 10^{-4}$ = grams of water vapor.

There is, in general, a similar trend in permeability to the two gases, and very large differences in oxygen permeability exist for some films that offer similar resistance to water vapor transfer.

Note that the units of permeability in Tables 10.2 and 10.3 differ markedly, as do the levels of relative humidity at which the measurements were made. There are, unfortunately, many different units in common use, but Selby (1961) has provided conversion factors for 12 of them.

The data in Tables 10.2 and 10.3 are given for films of the same thickness. Thicker films will usually have lower permeabilities, greater strength, and higher cost. In many food packaging applications, the materials are combined by lamination, coating, or coextrusion, and they may have properties very different from the basic or individual film. It is not appropriate here to discuss the many types of packaging materials now available. Instead, the requirements for packaging the various types of foods will be described. Comprehensive reviews of the technology of food packaging in films are provided by Selby (1961) and Brody (1970).

Microbiological problems may arise when moist foods are packaged in materials such as films coated with PVDC (a copolymer of vinyl and vinylidene chlorides), in glass or in metal. Water vapor transfer through these materials will be low or nonexistent. The packaged food will equilibrate with the internal atmosphere of the package, and subsequent cooling may cause moisture condensation on the internal surfaces of the package. The condensate creates localized areas of very high a_w in which microbial growth may occur. Thus, it is imperative that only foods of relatively low a_w be sealed into moisture-impermeable packages unless some supplementary scheme for microbial control is used. On the other hand, when high a_w foods are packaged in permeable materials, care must be taken to prevent excessive weight loss of the product. This may be achieved either by the careful selection of the wrapping materials or by control of humidity in the atmosphere surrounding the primary package.

UNREFRIGERATED PACKAGED FOODS

Dehydrated vegetables, which usually have levels of a_w of 0.30 or below, require protection against oxidation, as well as moisture uptake. Laminates of polyethylene, aluminum foil, and paper have been successfully used in packaging dehydrated vegetables. Dried fruits are prepared to much higher a_w levels and consequently, unless ambient humidities are very high, are safely stored in relatively permeable film. Protection against insect attack may be more important. The sensitivity of dried milk powder and egg powder to oxidation and to moisture absorption has frequently led to gas packaging, under nitrogen, in metal containers. However, for normal retail distribution, dried milk is best protected by paper–aluminum foil–polyethylene laminates. Dehydrated soups and other powdered products also require thorough protection from the ingress of oxygen and moisture, and heat-sealable packages of aluminum foil laminates are again routinely used.

Among the most susceptible to unsuitable storage conditions are those dry foods that depend for their appeal, in part, on their crispness. Typical of these are potato chips and roasted nuts. Additionally, these foods contain substantial amounts of lipids, and protection from uptake of both moisture and oxygen is again essential. Potato chips and similar snacks are very low-density products, so that the cost of packaging assumes additional importance. Small packs are commonly of PVDC-coated glassine and larger-sized packs of PVDC-coated cellulose or polypropylene. For long storage of nuts, vacuum packaging in cans or glass jars is desirable. Aluminum foil laminate pouches, again with vacuum packaging, are satisfactory, but, because retail turnover of small units is usually rapid, it is common practice to use cellulose–plastic combinations, which in addition to being cheaper, are transparent. Cookies and similar biscuits also have high fat contents and depend upon crispness for much of their appeal. They are successfully packaged in a wide range of materials with low W.V.T.R. and oxygen permeability.

REFRIGERATED PACKAGED FOODS

In the packaging of fresh meats for chilled storage, there are two conflicting requirements. To maintain the desirable red oxymyoglobin pigment, a film of high oxygen permeability is necessary, while to minimize the growth of spoilage bacteria, a relatively gas-impermeable film is appropriate to limit oxygen inflow and conserve within the package the inhibitory atmosphere of carbon dixoide formed by meat and microorganisms. In practice, a film highly permeable to oxygen is used for fresh meat to be stored only 2 or 3 days. Whether this film should permit high or low water vapor transmission rates will depend on whether fogging of the package is considered an important factor in the merchandising of the particular cut of meat.

For longer periods of chilled storage of fresh meat, films highly impermeable to both oxygen and moisture vapor and suitable for vacuum packaging are favored. Appropriate films usually rely on a layer of vinylidene chloride–vinyl chloride copolymer as the basic material. Although the anaerobic conditions imposed result in the formation of the dark myoglobin pigment, this is rapidly transformed to oxymyoglobin when the meat is exposed to air.

Similar types of vacuum packaging are used for chilled storage of cured meats to prevent dehydration, to protect the pigment (in this case, nitrosomyoglobin) from oxidation and to provide the anaerobic conditions that retard microbial spoilage.

For the frozen storage of meat, the deterioration of greatest concern is frequently "freezer burn," caused by severe desiccation of the surface layers coupled with oxidation. Film permeability is less important here than might be expected because both vapor penetration and reaction rates of oxygen with fats decrease with decreasing temperature and are very low in frozen storage. The main cause of freezer burn is migration of water from the product to the internal surface of the package as a consequence of fluctuations in temperature. Reduction in headspace volume reduces freezer burn. For low transmission of oxygen, moisture, and volatile flavor components, PVDC-coated films are very effective. For irregular-shaped items, such as poultry, or whenever minimum headspace volume is desired, more permeable but shrinkable films, such as polyethylene, may be preferred.

The optimum storage temperatures for the fresh fruit and vegetables described previously (Table 10.1) are much more readily obtained in an efficient cool store than within an individual package. However, since prepackaged items are held for relatively short periods, the conditions are less critical. A compromise is necessary between packaging materials that retain water and, hence maintain crispness, but cause condensation and fogging of the film, and those which permit weight loss and loss of crispness but, as a consequence, do not fog. Adequate gas permeability is required to prevent reduction of the oxygen concentration within the package to a level that may result in physiological disorders in the living tissues. Where gas permeability of the intact film is inadequate, perforations will admit oxygen without permitting excessive water loss—indeed, without necessarily overcoming a fogging problem, should one exist. However, fogging is commonly a temporary inconvenience, disappearing when the microdroplets coalesce.

For frozen vegetables, the most commonly used containers are polyethylene or paperboard, either waxed, plastic-coated or enclosed in a moisture-proof film. Similar packages, with special care taken to minimize oxygen uptake, are suitable for frozen fish. Frozen fruit juice concentrates are packaged in hermetically sealed or aluminum-laminated paperboard tin-plate composite containers of cans treated with appropriate lacquers. The use of an aluminum foil tray for cooked foods that contain liquids, such as sauce or gravy, provides some rigidity, as well as suitable moisture and oxygen barriers if the lid is adequately sealed to the tray.

The major defect developing during long-term storage and distribution of cheeses is surface mold growth, which is prevented largely by exclusion of oxygen. For processed and many soft cheeses, this is achieved by aluminum foil wraps and, for many others, by vacuum-packing in low-permeability plastic films.

This discussion of packaging is brief and confined largely to principles because of the speed with which this branch of food technology is developing. Detailed information on the properties and uses of packaging materials are given in the annual publication, "Modern Packaging Encyclopedia" (McGraw-Hill, New York).

REFERENCES

Brody, A. L. (1970). Flexible packaging of foods. *Crit. Rev. Food Technol.* **1,** 71–155.

Bulmer, C. H. (1965). The selection of packaging materials for foodstuffs. *Food Sci. Technol., Proc. Int. Congr., 1st, 1962* Vol. 4, pp. 363–368.

Davis, E. G. (1970). Evaluation and selection of flexible films for food packaging. *Food Technol. Aust.* **22,** 62–67.

Griffiths, H. J. (1967). Wet-bulb control of grain aeration systems. *CSIRO, Div. Mech. Eng., Circ.* No. 3.

Hall, E. G. (1973). Mixed storage of foodstuffs. *CSIRO Div. Food Res., Circ.* No. 9.

Karel, M. (1973). Quantitative analysis of food packaging and storage stability problems. *Chem. Eng. Prog. Symp. Ser.* **69,** 107–113.

Middlehurst, J. (1968). Condensation in uninsulated ISO containers. *Aust. Packag.* **16,** 24–27.

Mitchell, F. G., Guillou, L., and Parsons, R. A. (1972). Commercial cooling of fruit and vegetables. *Univ. Calif., Div. Agric. Sci., Manual* No. 43.

Niebergall, H. (1974). Der Einfluss der Gas- und Dampfdurchlassigkeit von Verpackungsfolien auf die Haltbarkeit folienverpackter Lebensmittel. *Chem., Mikrobiol., Technol. Lebensm.* **3,** 39–51.

Poisson, J., and Guilbot, A. (1963). Storage conditions and the length of grain storage. *Meun. Fr.* **193,** 19–29.

Scott, W. J. (1936). The growth of microorganisms on ox muscle. 1. The influence of water content of substrate on rate of growth at −1°C. *J. Counc. Sci. Ind. Res. (Aust.)* **9,** 177–190.

Selby, J. W. (1961). Modern food packaging film technology. *B.F.M.I.R.A. Sci. Tech. Surv.* No. 39.

Thompson, J. A., Sefcovic, M. S., and Kingsolver, C. H. (1962). Maintaining quality of pea beans during shipment overseas. *U.S. Dep. Agric., Mark. Res. Rep.* **519.**

11

Food Plant Sanitation

The subject of sanitation spans virtually all aspects of food harvesting, processing, packaging, and storage. It involves people, equipment, physical facilities, climate, geographical location, and many other factors. Given this broad involvement, it is not surprising that water activity governs some, if not many, of the actions that food sanitarians take to assure wholesome food supplies.

Consideration of a_w alone does not permit adequate coverage of this topic. In many cases, ambient relative humidity influences the a_w of foods during processing and storage, thus altering their a_w and affecting the strategy that must be used to control deteriorative effects. For this reason, a discussion of atmospheric relative humidity and its influence on sanitation and the protection of food products is included.

PROCESS EQUIPMENT CLEANING

Virtually all food plant cleaning and sanitizing agents currently in use are water soluble. In many cases, potable water rinses follow cleaning to assure that residues of detergents and/or sanitizers do not remain in the equipment to contaminate the product.

In many cleaning or sanitizing situations, such as dehydrating or dry-food processing plants, cleaning with water or water solutions may be counterproductive. That is, the use of water may actually hydrate food accumulations in process equipment or elsewhere, allowing bacteria to grow and/or providing food and harborage for insects. Before water is used for such purposes, a careful assessment of proposed cleaning programs should be conducted to assure that cleaning objectives can be met without jeopardizing the wholesomeness of the product. Many dry processes, of course, require water cleaning. However, in these cases, precleaning, preferably by vacuuming and thorough drying following wet cleaning, is mandatory. Drying can be accomplished by many methods, such as hot air, starting up spray or drum driers before product introduction, operating hot water or steam jackets before processing, etc. These operations are ideally conducted before equipment downtime, as before a weekend, to allow further drying under ambient conditions during the nonoperating period. The use of filtered, compressed air for such purposes is seldom effective, since this merely displaces material to other plant areas. The use of drying cloths should also be discouraged, as they often become heavily contaminated.

Throughout process cleaning operations, plant management and workers must be cognizant of the relationships between a_w and product quality and wholesomeness.

FOOD STORAGE SANITATION

Food normally is stored in large warehouses, many of them refrigerated, depending on the particular product requirements. Usually, some attempt is made to control the relative humidity in such storage facilities.

Ambient humidity may be measured and recorded continuously by mechanical hygrometers or dry/wet bulb hygrometers, such as sling or Assman hygrometers (Chapter 2), which may be used for this purpose. In fact, two parameters, relative humidity and temperature, are critical during storage of foods because the relation between them determines the dew point or temperature at which water vapor condenses from the air. Condensation can cause serious problems in a food storage warehouse when water permeation of packaging materials allows glues to lose their binding ability and the integrity of packages is jeopardized. Insects and other pests find humid conditions more to their liking than arid atmospheres, and microbial growth, especially of molds (see Chapter 10), is likely in damp surroundings. Warehouse operators must know that foods removed from cold or cool storage may ''sweat'' if the ambient humidity/temperature are such that the surface dew point of the stored material is exceeded. Subsequent transfer of these materials to a tightly closed rail car or truck may compound the problem, allowing molds to grow in transit. Usually, a period of equilibration is needed when conditions indicate that a condensation problem might develop.

EFFECT OF RELATIVE HUMIDITY ON INSECTS AND INSECT CONTROL

Many insects require only minute amounts of moisture for survival and growth; however, atmospheric humidity may greatly influence the extent to which insect control measures are effective in the food processing environment.

Beyond the obvious and direct means of insect control through moisture manipulation, such as the flooding of cranberry bogs to rid them of pests, or, alternatively, the drainage of marshes to eliminate mosquitoes, many insect control procedures are, to a greater or lesser degree, influenced by relative humidity and moisture.

Many insects are protected by a hard, waterproof exoskeleton or epicuticle, which owes its water impermeability to a lipoidal surface coating. One of the most effective means used by ancient man to rid his dwellings of insects was to sprinkle colonies with soot or road dust. Today, other materials such as powdered activated charcoal, magnesium, or calcium carbonate or finely powdered alumina serve a similar function which is to withdraw moisture from the interior of the insect (Brown, 1951) and, thus, kill it. In addition, many of these substances are abrasive. As a treated insect preens itself, lacerations of the cuticle may occur, effectively disrupting the lipoidal layer and allowing moisture to leave the insect. Death ensues as a result of dehydration.

Relative humidity and moisture may also play a part in determining the effectiveness of some insecticides. Gosswald (1934) found that the resistance of insect larvae to pyrethrum dusts increased as the relative humidity approached saturation.

EFFECT OF RELATIVE HUMIDITY ON FUMIGANTS AND ANTIMICROBIAL AGENTS

A number of foods are routinely treated with gaseous substances to reduce microbial contamination of their surfaces. Propylene oxide, ethylene oxide, and methyl bromide have been used for this purpose; regulations governing the use and concentrations of these materials vary from country to country.

Ethylene oxide may be used to reduce bacterial counts in ground spices, whereas propylene oxide has somewhat broader applicability. Among the foods that have been treated by the latter gas are glacé fruits, cocoa, and nut meats. Relative humidity and a_w seem to influence the effectiveness of all three gases mentioned above. Ethylene oxide acts optimally within a relative humidity range of 20–40% (Kaye and Phillips, 1949), probably as a result of the formation of cross-linkages or water molecule bridges between or within bacterial protein molecules. These bridges, reportedly, would make alkylating sites more accessible to ethylene oxide treatment and enhance its effectiveness (Gilbert *et al.*, 1964). While moisture appears to influence ethylene oxide potency, Kereluk *et al.* (1970) concluded that the a_w of the microenvironment influences the resistance of bacterial spores to this material to a greater degree. These authors suggested that moisture can become a critical factor when packaging materials prevent penetration into the package. Since some foodstuffs, such as ground spices, are normally treated with ethylene oxide as bulk commodities, these problems would not be encountered. Based on these and other data, the a_w of the treated material probably is at least as important as the actual relative humidity, since it is a_w which determines, at equilibrium, the a_w of the microenvironment in which the organisms exist.

There is disagreement as to whether the inhibitory action of propylene oxide increases or decreases with increasing atmospheric moisture. Bruch and Kosterer (1961) investigated the effect of propylene oxide vapor on bacteria in powdered and flake foods and observed that this chemical is more effective at low relative humidities. Conversely, Himmelfarb *et al.* (1962) found that *Bacillus subtilis* spores inoculated onto calcium alginate wool sponges are relatively less sensitive to propylene oxide at low relative humidity levels. Spore resistance decreased as the relative humidity of the test chambers was increased from 65 to 70%.

Others (Tawaratani and Shibasaki, 1972) have suggested that the moisture content of the spores (in this case, fungal spores) is more important in determining resistance than the relative humidity of the exposure atmosphere. These authors also noted that the amount of adsorbed propylene oxide, propylene glycol, and other residues increases with relative humidity. The treatment of shelled pecans (Beuchat, 1973) with propylene oxide appears to be more effective with increasing relative humidity, especially at higher concentrations (400 and 800 ppm). Beuchat suggested fumigation at relative humidity levels at or below the equilibrium relative humidity of the nut meats.

Based on these and other reports, it is difficult to suggest preferred relative humidity levels for the use of either ethylene or propylene oxides. Obviously, such factors as concentration, temperature (Beuchat, 1973), propensity for residue formation, moisture condition or a_w of the food, and the microenvironment surrounding the microorganisms, as well as the relative humidity conditions during treatment influence effectiveness.

Methyl bromide finds its principal application as a fumigant for the eradication of insects from certain stored, raw commodities and as a rail car or warehouse fumigant. Like ethylene and propylene oxides, this material is toxic to man, and appropriate precautions must be taken during its use (Mallis, 1969). Harry *et al.* (1973) have investigated the antibacterial activity of this compound under a wide range of conditions and have observed that its activity is adversely affected in high moisture (73%) materials. Some differences in susceptibility between different types of bacteria were also noted by these authors, with micrococci being significantly more resistant than either *Salmonella typhimurium* or *Escherichia coli*.

EFFECT OF RELATIVE HUMIDITY AND a_w ON BACTERIAL SURVIVAL ON SURFACES

The ability of bacteria to survive on surfaces within a food processing operation constitutes a critical area of concern for the food sanitarian. Survival on food contact surfaces is, of course, most important; however, survival on ceramic or synthetic floor materials, painted ceilings, and walls also may influence the total microbial "load" in food processing environments.

Under most circumstances, the a_w of surface films is less than that required for microbial growth. If it is not, the factors governing growth in wet systems, be they films, microbiological media, or foods, are covered elsewhere in this volume. Our primary concern, then, is survival, because the ability of microorganisms to maintain viability contributes to the eventual microbial populations in foods contacting these surfaces.

Surface films represent a moisture microenvironment that changes as water evaporates from the large surface areas normally exposed. The rate of evaporation is governed by ambient relative humidity, temperature, and the inherent a_w of the film. In the present context, the important factor is how long the a_w of the film is maintained in the range permitting microbial growth. Whether growth is possible will, of course, depend on the suitability of nutrients present in the film. Maxcy (1971) has listed the presence of soil and detergent residues, temperature, and population density as factors that influence the microflora on stainless steel surfaces. To this list might be added the physical nature of the surface, since Barnhart *et al.* (1970) have shown that rough, corroded surfaces provide a more favorable microenvironment for microbial growth in milk films. These workers also showed that relative humidity levels of 12, 50, and 80% differed little in their ability to sustain bacterial populations in the films. In contrast, Maxcy (1971) found that the a_w of the microenvironment is a major factor controlling the "density" of microorganisms in films on stainless steel surfaces. Also, the a_w of the film has a selective effect on the type of bacteria present. At 0.80 a_w, the majority of bacteria surviving were gram-positive organisms; whereas, at an a_w of 0.99, more than 50% of the organisms present were gram negative. Data on the survival of *Staphylococcus aureus* on a stainless steel surface is presented in the publication of McDade and Hall (1963) and in Fig. 11.1.

The basic task of any surface-cleaning regimen in a food plant is to remove food residues and microorganisms. If the cleaning operation is thorough, these surfaces should be relatively inhospitable to the survival of any remaining microorganisms. The ambient relative humidity appears to influence microbial survival by governing the rate and extent of surface film drying. Other factors, as noted above, are also involved. To achieve minimal retention of such films and whatever bacterial numbers they may harbor, surfaces should be smooth, rapid

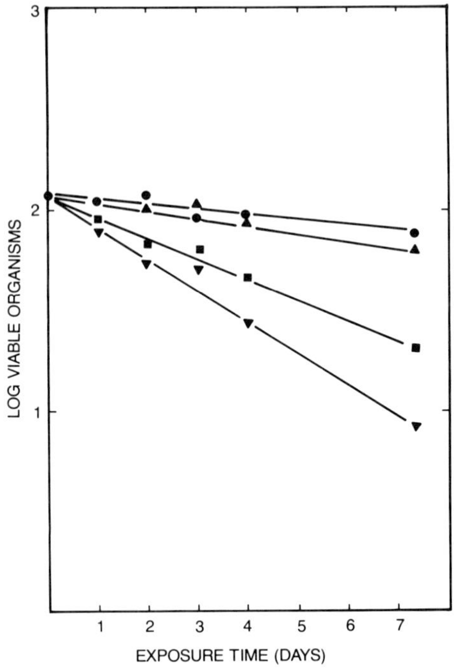

Fig. 11.1. Survival of *Staphylococcus aureus* FDA 209 on polished stainless steel surfaces. Relative humidities are (●) 11%, (▲) 33%, (■) 53%, (▼) 85%. Temperature is 25°C (from McDade and Hall, 1963).

and thorough air drying should be achieved, and a reduced atmospheric humidity maintained in the plant.

The proper handling of very humid air emanating from food processes is one of the most difficult problems facing the food sanitarian. If not properly contained, such air usually rises and eventually encounters a cooler surface, whereupon the dew point is exceeded and moisture condensation occurs. The result is that overhead painted surfaces may flake, creating product hazards. Also, food dusts may be wetted to the point that mold and even bacterial growth may occur, and pipe insulation may absorb sufficient moisture to support mold growth. In addition, condensed water can drop into process areas, creating sanitation and safety hazards.

Although various measures may be employed to reduce the effects of such moisture, it is best eliminated at the source through proper venting to the outside atmosphere. Where this is practical, vent lines must be unobstructed, short, and straight. Exhaust systems such as this should be considered a part of the process and must be designed to assure easy cleaning.

EFFECT OF RELATIVE HUMIDITY ON AIRBORNE MICROORGANISMS

The airborne transfer of microorganisms may constitute a significant source of contamination in many food-processing plants. Although much of the information on airborne contamination that has developed is a result of investigations performed in hospital environments by McDade and Hall (1963), many of their observations apply to food processing establishments and may be instructive to the food sanitarian attempting to prevent product contamination from airborne sources.

Airborne contamination is usually composed of particulate material, either liquid or solid, plus whatever microorganisms are carried by these particles. In the case of aerosolized sputum or spattering from floor-washing operations, the microbial content might be quite high. Conversely, fewer organisms might be found on dust derived from the grinding of steel. The influence of relative humidity on the persistence of microorganisms on or in airborne particles can be a key factor in determining the microbial contamination of a food product or process. Dunklin and Puck (1948) investigated the survival, at various relative humidities, of three bacterial species atomized into experimental chambers.

They found that maximal kill occurs at 50% relative humidity if the organisms were initially suspended on physiological saline or sputum. Cell death was found to occur during evaporation of the matrix. If water is the suspending medium, an optimum humidity for microbial kill is not observed.

Lidwell and Lowbury (1950) found a positive correlation between atmospheric relative humidity and the death rate of strains of *Staphylococcus aureus* and β-hemolytic streptococci. These authors, however, could not rule out the possibility that a contrary correlation might exist for gram-negative organisms such as *E. coli*. The possibility was also raised that variations in the free-water content of the matrix (e.g., saliva or dust) might account for death rate differences.

Conversely, Lighthart *et al.* (1971) described a reduction in the death rate of airborne *Serratia marcescens* cells with an increase in relative humidity to 80%. Riley and Kaufman (1972) noted a similar effect with this organism in atmospheres exposed to UV radiation. The sharpest declines in death rates were noted by these authors as the relative humidity levels were increased from 50 to 70%, followed by reactivation at humidities > 80% at comparable radiation doses. Although confirming that bacterial resistance to ultraviolet and higher wavelengths increases at relative humidity levels > 70%, Webb (1965) found that the opposite was true for X-irradiation. This difference was speculated to be the result of the dependence of large conjugate nucleoproteins on correctly oriented bound water molecules for their stability. Because X-rays "attack" these molecules directly, the destructive effects of this irradiation would be felt keenly at high humidities.

Ultraviolet, on the other hand, requires energy to displace water, and so with greater amounts of water present or easily replaceable, this process becomes increasingly less efficient at higher moisture levels.

Completing the spectrum of effects noted, Ehrlich *et al.* (1970) found that the survival of several species of *Flavobacterium* was not significantly affected by relative humidity levels ranging from 25 to 99%. Similar effects were observed for aerosol mixtures of three different viruses by Mayhew and Hahon (1970). Biological decay rates were not affected significantly by relative humidity levels of 30, 50, and 80%.

One of the most complete reviews of the influence of relative humidity on microbial survival was published by Webb (1965). This author stated that the death rate of cells atomized in a special chamber is dependent on the cell-bound water content at the time that equilibrium is reached with the environmental water vapor. Webb, like others (Wells and Wells, 1936), found that high relative humidities are more protective than low humidities; however, certain carbohydrates such as glucose, glucosamine, and inositol tend to circumvent the destructive effects on bacterial aerosols at low relative humidity. This effect was hypothesized to be the result of water molecule replacement by compounds, such as the above, which are capable of forming hydrogen bonds. Evidence that this is the case was shown by the fact that genera highly resistant to drying effects, such as *Mycobacterium* and *Staphylococcus,* possess unusually high intracellular inositol levels.

These conflicting data make conclusive interpretation of relative humidity effects on airborne contamination very difficult. Obviously, further work is needed, using standardized aerosols, before definite conclusions can be stated. While the fate of airborne particles and microorganisms is of consequence to food sanitarians, the primary emphasis should be placed on good hygiene and cleaning practices that minimize or prevent the formation of aerosols, prevent their entrance to food product streams, and remove them from the air once an aerosol of dirt and dust has formed.

REFERENCES

Barnhart, H. M., Jr., Maxcy, R. B., and Georgi, C. E. (1970). Effect of humidity on the fate of the microflora of milk on stainless steel surfaces. *Food Technol.* **24,** 1385–1389.

Beuchat, L. R. (1973). *Escherichia coli* on pecans: Survival under various storage conditions and disinfection with propylene oxide. *J. Food Sci.* **38,** 1064–1066.

Brown, A. W. A. (1951). "Insect Control by Chemicals." Wiley, New York.

Bruch, C. W., and Kosterer, M. G. (1961). The microbiocidal activity of gaseous propylene oxide and its application to powdered or flaked foods. *J. Food Sci.* **26,** 428–435.

Dunklin, E. W., and Puck, T. T. (1948). The lethal effect of relative humidity on airborne bacteria. *J. Exp. Med.* **87,** 87–101.

Ehrlich, R., Miller, S., and Walker, R. L. (1970). Effects of atmospheric humidity and temperature on the survival of airborne Flavobacterium. *Appl. Microbiol.* **20,** 884–887.

Gilbert, G. L., Gambil, V. M., Spinner, O. R., Hoffman, R. K., and Phillips, C. R. (1964). Effect of moisture on ethylene oxide sterilization. *Appl. Microbiol.* **12,** 496–503.

Gosswald, K. (1934). Temperature optimum and resistance to insecticides. *Z. Angew. Entomol.* **20,** 489–530.

Harry, E. G., Brown, W. B., and Goodship, G. (1973). The influence of temperature and moisture on the disinfecting activity of methyl bromide on infected poultry litter. *J. Appl. Bacteriol.* **36,** 343–350.

Himmelfarb, P. H., El-Bisi, M., Read, R. B., Jr., and Litsky, W. (1962). Effect of relative humidity on the bactericidal activity of propylene oxide vapor. *Appl. Microbiol.* **10,** 431–435.

Kaye, S., and Phillips, C. R. (1949). The sterilizing action of gaseous ethylene oxide. IV. The effect of moisture. *Am. J. Hyg.* **50,** 296–306.

Kereluk, K., Gammon, R. A., and Lloyd, R. S. (1970). Microbiological aspects of ethylene oxide sterilization. III. Effects of humidity and water activity on the sporicidal activity of ethylene oxide. *Appl. Microbiol.* **19,** 157–162.

Lidwell, O. M., and Lowbury, E. J. (1950). The survival of bacteria in dust. II. The effect of atmospheric humidity on the survival of bacteria in dust. *J. Hyg.* **48,** 21–27.

Lighthart, B., Hiatt, V. E., and Rossano, A. T., Jr. (1971). The survival of airborne Serratia marcescens in urban concentration of sulfur dioxide. *J. Air Pollut. Control Assoc.* **21,** 639–642.

McDade, J. V., and Hall, L. B. (1963). Survival of *Staphylococcus aureus* in the environment. I. Exposure on surfaces. *Am. J. Hyg.* **78,** 330–337.

Mallis, A. (1969). "Handbook of Pest Control." MacNair-Dorland Co., New York.

Maxcy, R. B. (1971). Factors in the ecosystem of food processing equipment contributing to outgrowth of microorganisms on stainless steel surfaces. *J. Milk Food Technol.* **34,** 569–573.

Mayhew, C. J., and Hahon, N. (1970). Assessment of aerosol mixtures of different viruses. *Appl. Microbiol.* **20,** 313–316.

Riley, R. L., and Kaufman, J. E. (1972). Effect of relative humidity on the inactivation of airborne *Serratia marcescens* by ultraviolet radiation. *Appl. Microbiol.* **23,** 1113–1120.

Tawarantani, T., and Shibasaki, I. (1972). Effect of moisture content on the microbiocidal activity of propylene oxide and the residue in foodstuffs. *J. Ferment. Technol.* **50,** 349–353.

Webb, S. J. (1965). "Bound Water in Biological Integrity." Thomas, Springfield, Illinois.

Wells, W. F., and Wells, N. W. (1936). Air-borne infection. *J. Am. Med. Assoc.* **107,** 1698–1703.

Appendix A

Approximate a_w Values of Some Foods and of Sodium Chloride and Sucrose Solutions

a_w	NaCl (%)	Sucrose (%)	Foods
1.00–0.95	0–8	0–44	Fresh meat, fruit, vegetables, canned fruit in syrup, canned vegetables in brine, frankfurters, liver sausage, margarine, butter, low-salt bacon
0.95–0.90	8–14	44–59	Processed cheese, bakery goods, high-moisture prunes, raw ham, dry sausage, high-salt bacon, orange juice concentrate
0.90–0.80	14–19	59-saturation (0.86 a_w)	Aged cheddar cheese, sweetened condensed milk, Hungarian salami, jams, candied peel, margarine
0.80–0.70	19-saturation (0.75 a_w)	—	Molasses, soft dried figs, heavily salted fish
0.70–0.60	—	—	Parmesan cheese, dried fruit, corn syrup, licorice
0.60–0.50	—	—	Chocolate, confectionery, honey, noodles
0.40	—	—	Dried egg, cocoa
0.30	—	—	Dried potato flakes, potato crisps, crackers, cake mixes, pecan halves
0.20	—	—	Dried milk, dried vegetables, chopped walnuts

Appendix B

Approximate Minimum Levels of a_w Permitting Growth of Microorganisms at Temperatures Near Optimal

Molds	a_w
Alternaria citri	0.84
Aspergillus candidus	0.75
A. conicus	0.70
A. flavus	0.78
A. fumigatus	0.82
A. niger	0.77
A. ochraceous	0.77
A. restrictus	0.75
A. sydowii	0.78
A. tamarii	0.78
A. terreus	0.78
A. versicolor	0.78
A. wentii	0.84
Botrytis cinerea	0.93
Chrysosporium fastidium	0.69
C. xerophilum	0.71
Emericella (Aspergillus) nidulans	0.78
Eremascus albus	0.70
E. fertilis	0.77
Eurotium (Aspergillus) amstelodami	0.70
E. carnoyi	0.74
E. chevalieri	0.71
E. echinulatum	0.62
E. herbariorum	0.74
E. repens	0.71
E. rubrum	0.70
Monascus (Xeromyces) bisporus	0.61
Mucor plumbeus	0.93

Molds	a_w
Paecilomyces variotii	0.84
Penicillium brevicompactum	0.81
P. chrysogenum	0.79
P. citrinum	0.80
P. cyclopium	0.81
P. expansum	0.83
P. fellutanum	0.80
P. frequentans	0.81
P. islandicum	0.83
P. martensii	0.79
P. palitans	0.83
P. patulum	0.81
P. puberulum	0.81
P. spinulosum	0.80
P. viridicatum	0.81
Rhizopus nigricans	0.93
Rhizoctonia solani	0.96
Stachybotrys atra	0.94
Wallemia sebi (Sporendonema epizoum)	0.75
Yeasts	a_w
Debaryomyces hansenii	0.83
Saccharomyces bailii	0.80
S. cerevisiae	0.90
S. rouxii	0.62
Bacteria	a_w (adjusted with salts)
Aerobacter aerogenes	0.94
Bacillus cereus	0.95
B. megaterium	0.95
B. stearothermophilus	0.93
B. subtilis	0.90
Clostridium botulinum type A	0.95
C. botulinum type B	0.94
C. botulinum type E	0.97
C. perfringens	0.95
Escherichia coli	0.95
Halobacterium halobium	0.75
Lactobacillus viridescens	0.95
Microbacterium spp.	0.94
Micrococcus halodenitrificans	0.86
M. lysodeikticus	0.93
Pseudomonas fluorescens	0.97
Salmonella spp.	0.95
Staphylococcus aureus	0.86
Vibrio costicolus	0.86
V. parahaemolyticus	0.94

Index

G

N

R

S

X

Y

Z

A
B
C 8
D 9
E 0
F 1
G 2
H 3
I 4
J 5